A Proof of Alon's
Second Eigenvalue Conjecture
and Related Problems

A Proof of Alon's
Second Eigenvalue Conjecture
and Related Problems

MEMOIRS
of the
American Mathematical Society

Number 910

A Proof of Alon's
Second Eigenvalue Conjecture
and Related Problems

Joel Friedman

2000 *Mathematics Subject Classification.* Primary 68R10, 05C50.

Library of Congress Cataloging-in-Publication Data

Friedman, Joel, 1962–
  A proof of Alon's second eigenvalue conjecture and related problems / Joel Friedman.
    p. cm. — (Memoirs of the American Mathematical Society, ISSN 0065-9266 ; no. 910)
  Includes bibliographical references.
  ISBN 978-0-8218-4280-5 (alk. paper)
  1. Eigenvalues. I. Title.
QA193.F75   2008
512.9′436—dc22   2008020746

Memoirs of the American Mathematical Society

This journal is devoted entirely to research in pure and applied mathematics.

Subscription information. The 2008 subscription begins with volume 191 and consists of six mailings, each containing one or more numbers. Subscription prices for 2008 are US$675 list, US$540 institutional member. A late charge of 10% of the subscription price will be imposed on orders received from nonmembers after January 1 of the subscription year. Subscribers outside the United States and India must pay a postage surcharge of US$38; subscribers in India must pay a postage surcharge of US$43. Expedited delivery to destinations in North America US$53; elsewhere US$130. Each number may be ordered separately; *please specify number* when ordering an individual number. For prices and titles of recently released numbers, see the New Publications sections of the *Notices of the American Mathematical Society*.

Back number information. For back issues see the *AMS Catalog of Publications*.

Subscriptions and orders should be addressed to the American Mathematical Society, P. O. Box 845904, Boston, MA 02284-5904, USA. *All orders must be accompanied by payment.* Other correspondence should be addressed to 201 Charles Street, Providence, RI 02904-2294, USA.

Copying and reprinting. Individual readers of this publication, and nonprofit libraries acting for them, are permitted to make fair use of the material, such as to copy a chapter for use in teaching or research. Permission is granted to quote brief passages from this publication in reviews, provided the customary acknowledgment of the source is given.

Republication, systematic copying, or multiple reproduction of any material in this publication is permitted only under license from the American Mathematical Society. Requests for such permission should be addressed to the Acquisitions Department, American Mathematical Society, 201 Charles Street, Providence, Rhode Island 02904-2294, USA. Requests can also be made by e-mail to `reprint-permission@ams.org`.

Memoirs of the American Mathematical Society (ISSN 0065-9266) is published bimonthly (each volume consisting usually of more than one number) by the American Mathematical Society at 201 Charles Street, Providence, RI 02904-2294, USA. Periodicals postage paid at Providence, RI. Postmaster: Send address changes to Memoirs, American Mathematical Society, 201 Charles Street, Providence, RI 02904-2294, USA.

© 2008 by the American Mathematical Society. All rights reserved.
Copyright of this publication reverts to the public domain 28 years
after publication. Contact the AMS for copyright status.
This publication is indexed in *Science Citation Index*®, *SciSearch*®, *Research Alert*®,
CompuMath Citation Index®, *Current Contents*®/*Physical, Chemical & Earth Sciences*.
Printed in the United States of America.

∞ The paper used in this book is acid-free and falls within the guidelines
established to ensure permanence and durability.
Visit the AMS home page at `http://www.ams.org/`

10 9 8 7 6 5 4 3 2 1   13 12 11 10 09 08

Contents

Chapter 1. Introduction	1
Chapter 2. Problems with the Standard Trace Method	7
1. The Trace Method	7
2. Limitations of the Trace Expansion	12
Chapter 3. Background and Terminology	17
1. Graph Terminology	17
2. Variable-Length Graphs and Subdivisions	18
3. λ_1 of a VLG	19
4. Shannon's Algorithm and Formal Series	19
5. Limiting Graphs	22
6. Irreducible Eigenvalues	22
7. λ_1 and Closed Walks for Infinite Graphs	24
8. A Curious Theorem	24
Chapter 4. Tangles	27
Chapter 5. Walk Sums and New Types	33
1. Walk sums	34
2. The Loop	39
3. Forms, Types, and New Types	40
4. Motivation of Types and New Types	43
Chapter 6. The Selective Trace	47
1. The General Selective Trace	47
2. A Lemma on Selective Walks	47
3. Determining τ_{fund} for $\mathcal{G}_{n,d}$	51
4. Determining τ_{fund} for $\mathcal{H}_{n,d}$, $\mathcal{I}_{n,d}$, and $\mathcal{J}_{n,d}$	51
Chapter 7. Ramanujan Functions	57
Chapter 8. An Expansion for Some Selective Traces	59
Chapter 9. Selective Traces In Graphs With (Without) Tangles	65
Chapter 10. Strongly Irreducible Traces	73
Chapter 11. A Sidestepping Lemma	77
Chapter 12. Magnification Theorems	81
Chapter 13. Finishing the $\mathcal{G}_{n,d}$ Proof	87

Chapter 14.	Finishing the Proofs of the Main Theorems	91
Chapter 15.	Closing Remarks	95
Glossary		97
Bibliography		99

Abstract

A d-regular graph has largest or first (adjacency matrix) eigenvalue $\lambda_1 = d$. Consider for an even $d \geq 4$, a random d-regular graph model formed from $d/2$ uniform, independent permutations on $\{1,\ldots,n\}$. We shall show that for any $\epsilon > 0$ we have all eigenvalues aside from $\lambda_1 = d$ are bounded by $2\sqrt{d-1} + \epsilon$ with probability $1 - O(n^{-\tau})$, where $\tau = \lceil (\sqrt{d-1} + 1)/2 \rceil - 1$. We also show that this probability is at most $1 - c/n^{\tau'}$, for a constant c and a τ' that is either τ or $\tau + 1$ ("more often" τ than $\tau+1$). We prove related theorems for other models of random graphs, including models with d odd.

These theorems resolve the conjecture of Alon, that says that for any $\epsilon > 0$ and d, the second largest eigenvalue of "most" random d-regular graphs are at most $2\sqrt{d-1} + \epsilon$ (Alon did not specify precisely what "most" should mean or what model of random graph one should take).

Received by the editor March 27, 2002, and in revised form April 30, 2004.
1991 *Mathematics Subject Classification*. Primary 68R10, 05C50.
Key words and phrases. second eigenvalue, random, graph, regular, Alon conjecture.
Research supported in part by an NSERC grant.

CHAPTER 1

Introduction

The eigenvalues of the adjacency matrix of a finite undirected graph, G, are real and hence can be ordered
$$\lambda_1(G) \geq \lambda_2(G) \geq \cdots \geq \lambda_n(G),$$
where n is the number of vertices in G. If G is d-regular, i.e., each vertex is of degree d, then $\lambda_1 = d$. In [**Alo86**], Noga Alon conjectured that for any $d \geq 3$ and $\epsilon > 0$, $\lambda_2(G) \leq 2\sqrt{d-1} + \epsilon$ for "most" d-regular graphs on a sufficiently large number of vertices. The Alon-Boppana bound shows that the constant $2\sqrt{d-1}$ cannot be improved upon (see [**Alo86, Nil91, Fri93**]). The main goal of this paper is to prove this conjecture for various models of a "random d-regular graph."

Our methods actually show that for "most" d-regular graphs, $|\lambda_i(G)| \leq 2\sqrt{d-1} + \epsilon$ for all $i \geq 2$, since our methods are variants of the standard "trace method."

Our primary interest in Alon's conjecture, which was Alon's motivation, is that fact that graphs with $|\lambda_i|$ small for $i \geq 2$ have various nice properties, including being expanders or magnifiers (see [**Alo86**]).

For a fixed n we can generate a random d-regular graph on n vertices as follows, assuming d is even (later we will give random graph models that allow d to be even or odd). Take $d/2$ permutations on $V = \{1, \ldots, n\}$, $\pi_1, \ldots, \pi_{d/2}$, each π_i chosen uniformly among all $n!$ permutations with all the π_i independent. We then form
$$E = \left\{ \big(i, \pi_j(i)\big), \big(i, \pi_j^{-1}(i)\big) \mid j = 1, \ldots, d/2, \quad i = 1, \ldots, n \right\},$$
yielding a directed graph $G = (V, E)$, which we may view as undirected. We call this probability space of random graphs $\mathcal{G}_{n,d}$. G can have multiple edges and self-loops, and each self-loop contributes 2 to the appropriate diagonal entry of G's adjacency matrix[1].

The main goal of this paper is to prove theorems like the following, which prove Alon's conjecture, for various models of a random d-regular graph; we start with the model $\mathcal{G}_{n,d}$.

THEOREM 1.1. *Fix a real $\epsilon > 0$ and an even positive integer d. Then there is a constant, c, such that for a random graph, G, in $\mathcal{G}_{n,d}$ we have that with probability at least $1 - c/n^\tau$ we have for all $i > 1$*
$$|\lambda_i(G)| \leq 2\sqrt{d-1} + \epsilon,$$
where $\tau = \tau_{\text{fund}} = \lceil (\sqrt{d-1} + 1)/2 \rceil - 1$. Furthermore, for some constant $c' > 0$ we have that $\lambda_2(G) > 2\sqrt{d-1}$ with probability at least c'/n^s, where $s = \lfloor (\sqrt{d-1} +$

---

[1]Such a self-loop is a *whole-loop* in the sense of [**Fri93**]; see also Chapter 2 of this paper.

$1)/2\rfloor$. (So $s = \tau_{\text{fund}}$ unless $\left(\sqrt{d-1}+1\right)/2$ is an integer, in which case $\tau_{\text{fund}} = s - 1$.)

Left open is the question of whether or not this theorem holds with $\epsilon = 0$ (which would yield "Ramanujan graphs") or even some function $\epsilon = \epsilon(n) < 0$. Calculations such as those in [**Fri93**] suggest that it does, even for some negative function $\epsilon(n)$. Examples of "Ramanujan graphs," i.e., graphs where $|\lambda_i(G)| \le 2\sqrt{d-1}$ except $i = 1$ (and, at times, $i = n$ when $\lambda_n = -d$) have been given in [**LPS88, Mar88, Mor94**] where d is one more than an odd prime or prime power. Theorem 1.1 demonstrates the existence of "nearly Ramanujan" graphs of any even degree. We shall soon address odd d, as well.

Another interesting question arises in the gap between τ_{fund} and s in Theorem 1.1 in the case where $\left(\sqrt{d-1}+1\right)/2$ is an integer; it is almost certain that one of them can be improved upon. In the language of Chapter 4 of this paper, τ_{fund} is the smallest order of a supercritical tangle, and s that of a hypercritical tangle; a gap between τ_{fund} and s can only occur when there is a critical tangle of order smaller than that of any hypercritical tangle.

Previous bounds of the form $\lambda_2 \le f(d) + \epsilon$ include $f(d) = (2d)^{1/2}(d-1)^{1/4}$ of the author (see [**Fri03**]), which is slight improvement over the Broder-Shamir bound of $f(d) = 2^{1/2}d^{3/4}$ (see [**BS87**]). Asymptotically in d, the bounds $f(d) = C\sqrt{d}$ of Kahn and Szemerédi (see [**FKS89**], here C is some constant) and $f(d) = 2\sqrt{d-1} + 2\log d + C$ of the author (see [**FKS89, Fri91**] and see equation (6) for the more precise bound) are improvements over the first two bounds.

The value of τ_{fund} in Theorem 1.1 depends on the particular model of a random graph. Indeed, consider the model $\mathcal{H}_{n,d}$ of a random graph, which is like $\mathcal{G}_{n,d}$ except that we insist that each π_i be one of the $(n-1)!$ permutations whose cyclic decomposition consists of one cycle of length n. The same methods used to prove Theorem 1.1 will show the following variant.

THEOREM 1.2. *Theorem 1.1 holds with $\mathcal{G}_{n,d}$ replaced by $\mathcal{H}_{n,d}$ and $\tau_{\text{fund}} = \lceil\sqrt{d-1}\rceil - 1$ and $s = \lfloor\sqrt{d-1}\rfloor$, except that when $d = 4$ we take $s = 2$.*

Once again, $\tau_{\text{fund}} = s$, unless a certain expression, in this case $\sqrt{d-1}$ (excepting $d = 4$), is an integer. Note that for $\mathcal{H}_{n,d}$, the value of τ_{fund} is roughly twice as large as that for $\mathcal{G}_{n,d}$ for d large.

Next consider two more models of random d-regular graphs; in these two models d may be even or odd. Let $\mathcal{I}_{n,d}$, for positive integers n, d with n even, be the model of a random d-regular graph formed from d random perfect matchings on $\{1, \ldots, n\}$.

For an odd positive integer n, let a *near perfect matching* be a matching of $n-1$ elements of $\{1, \ldots, n\}$; such a matching becomes a 1-regular graph if it is complemented by a single half-loop[2] at the unmatched vertex. Taking d independent such 1-regular graphs gives a model, $\mathcal{J}_{n,d}$, of a d-regular graph on n vertices for n odd.

THEOREM 1.3. *Theorem 1.2 holds with $\mathcal{H}_{n,d}$ replaced by $\mathcal{I}_{n,d}$ and with no $d = 4$ exception (i.e., $s = 1$ for $d = 4$). Theorem 1.1 holds with $\mathcal{G}_{n,d}$ replaced by $\mathcal{J}_{n,d}$.*

We can assert the truth of the Alon conjecture on more models of random graphs by using results on contiguity and related notions. Consider two families of probability spaces, $(\Omega_n, \mathcal{F}_n, \mu_n)_{n=1,2,\ldots}$ and $(\Omega_n, \mathcal{F}_n, \nu_n)_{n=1,2,\ldots}$ over the same

---

[2]Readers unfamiliar with half-loops (i.e., self-loops contributing only 1 to a diagonal entry of the adjacency matrix) can see Chapter 2 of this paper or [**Fri93**].

sets Ω_n and sigma-algebras $\mathcal{F}_n$; denote $\mu = \{\mu_n\}$ and $\nu = \{\nu_n\}$. We say that μ *dominates* ν if for any family of measurable events, $\{E_n\}$ (i.e., $E_n \in \mathcal{F}_n$), we have $\mu_n(E_n) \to 0$ as $n \to \infty$ implies $\nu_n(E_n) \to 0$ as $n \to \infty$. We say that μ and ν are *contiguous* if μ dominates ν and ν dominates μ.

COROLLARY 1.4. *Fix an $\epsilon > 0$ and an integer $d \geq 2$. Let $\mathcal{L}_n$ be any family of probability spaces of d-regular graphs on n vertices (possibly defined for only certain n) that is dominated by $\mathcal{G}_{n,d}$, $\mathcal{H}_{n,d}$, $\mathcal{I}_{n,d}$, or $\mathcal{J}_{n,d}$. Then for G in $\mathcal{L}_n$ we have that with probability $1 - o(1)$ (as $n \to \infty$) for all i with $2 \leq i \leq n$ we have*
$$|\lambda_i(G)| \leq 2\sqrt{d-1} + \epsilon.$$

There are a lot of results regarding contiguity and (at least implicitly) domination; see [**GJKW02, KW01, Wor99**] and the references there. For example, if $\mathcal{G}'_{n,d}$ is the restriction of $\mathcal{G}_{n,d}$ to those graphs without self-loops, then for $d \geq 4$ it is known that (1) $\mathcal{G}'_{n,d}$ and $\mathcal{H}_{n,d}$ are contiguous (by [**KW01**] and previous work), and (2) $\mathcal{G}_{n,d}$ dominates $\mathcal{G}'_{n,d}$ (easy, since a self-loop occurs in $\mathcal{G}_{n,d}$ with probability bounded away from 1 for fixed d). Thus the Alon conjecture for $\mathcal{G}_{n,d}$ implies the same for $\mathcal{H}_{n,d}$ (but this contiguity and/or domination approach does not give as tight a bound on the probability that $\lambda_2 \leq 2\sqrt{d-1} + \epsilon$ fails to hold as is given in Theorem 1.2). Also, $\mathcal{G}_{n,d}$ is contiguous with the "pairing" or "configuration" model of d-regular (pseudo)graphs (see [**GJKW02**]); it follows that the Alon conjecture holds for the latter model, and thus (see [**Wor99**], especially the beginning of Section 2 and Corollary 4.17) the conjecture holds for n (and d) even for $\mathcal{I}_{n,d}$ or the uniform measure on all d-regular (simple) graphs on n vertices.

Our method for proving Theorems 1.1, 1.2, and 1.3 is a variant of the well-known "trace method" (see, for example [**Wig55, Gem80, FK81, BS87, Fri91**]) originated by Wigner, especially the author's refinement in [**Fri91**] of the beautiful Broder-Shamir style of analysis in [**BS87**]. The standard trace method involves taking the expected value of the trace of a reasonably high power[3] of the adjacency matrix. In our situation we are unable to analyze this trace accurately enough to prove Theorem 1.1, as certain infinite sums involved in our analysis diverge (for example, the infinite sum involving $W(T; \vec{m})$ and $P_{i,T,\vec{m}}$ just above the middle of page 351 in [**Fri91**], for types of order $> d$). This divergence is due to certain "tangles" that can occur in a random graph and can adversely affect the eigenvalues (see Chapters 2 and 4). To get around these "tangles" we introduce a *selective trace*. We briefly sketch what a selective trace is in the next paragraph.

Recall that a closed walk about a vertex, v, is a walk in the graph beginning and ending at v. Recall that the trace of the k-th power of the adjacency matrix equals the sum over all v of the number of closed walks about v of length k. The k-*th irreducible trace* (used in both [**BS87**] and [**Fri91**]) is the same sum as the k-th power trace, except that we require the closed walks to be *irreducible*, i.e., to have no edge traversed and then immediately thereafter traversed in the opposite direction. A selective trace is a sum like an irreducible trace, but where we further require that the walk have no small contiguous piece that "traces out" a "supercritical tangle" (the notions of "tracing out" and "supercritical tangles" will be defined later; roughly speaking, a "supercritical tangle" is a small graph with many cycles). Since these "tangles" occur with probability at most proportional to $n^{-\tau}$, with

---

[3]In [**BS87, Fri91**] this power is roughly $c \log n$, where c depends on d and on aspects of the method.

$\tau = \tau_{\text{fund}}$ as in Theorem 1.1, the selective trace usually agrees with the standard "irreducible" trace.

Analyzing the selective traces involves a new concept of the "new type," which is a refinement of the "type" of [**Fri91**].

We caution the reader about the notation used here. In this paper we work with only d-regular graphs. In [**BS87**] $2d$-regular graphs were studied; in [**Fri91**] the graphs are usually $2d$-regular, although for a part of Chapter 3 the graphs are d-regular. We also caution the reader that here we use the term "irreducible" (as used in [**BS87, Fri91**] and, for example, in the text [**God93**]) as opposed to "reduced" (which is quite common) or "non-backtracking" (sometimes used in [**Fri91**]) in describing walks and related concepts.

We hope to generalize or "relativize" the theorems here to theorems about new eigenvalues of random covers (see [**Fri03**] for a relativized Broder-Shamir theorem). In this paper we occasionally go out of our way to use a technique that will easily generalize to this setting.

The rest of this paper is organized as follows. In Chapter 2 we review the trace method used in [**Fri91**] and explain why it requires modification to prove Alon's conjecture; as a byproduct we establish the part of Theorem 1.1 involving s. In Chapter 3 we give some background needed for some technical details in later sections. In Chapter 4 we formalize the notion of a tangle, and discuss their properties; we prove the part of Theorem 1.2 and 1.3 involving s. In Chapter 5 we describe "types" and "new types," explaining how they help to estimate "walk sums;" walk sums are generalizations of all notions of "trace" used here. In Chapter 6 we describe the "selective trace" used in this paper; we give a crucial lemma that counts certain types of selective closed walks in a graph. In Chapter 7 we explain a little about "d-Ramanujan" functions, giving a theorem to be used in Chapter 14 that also illustrates one of the main technical points in Chapter 8. In Chapter 8 we prove that certain selective traces have an asymptotic expansion (in $1/n$) whose coefficients are "d-Ramanujan." In Chapter 9 we show that the expansion in Chapter 8 still exists when we count selective traces of graphs not containing any finite set of tangles of order ≥ 1. In Chapter 10 we introduce strongly irreducible traces, that simplify the proofs of the main theorems in this paper. In Chapter 11 we prove a crucial lemma that allows us to use the asymptotic expansion to make conclusions about certain eigenvalues; this lemma sidesteps the unresolved problem of (even roughly) determining the coefficients of the asymptotic expansion (in [**Fri91**] we actually roughly determine the coefficients for the shorter expansion developed there). In Chapter 12 we prove the magnification (or "expansion") properties needed to apply the sidestepping lemma of Chapter 11. In Chapter 13 we complete the proof of Theorem 1.1. In Chapter 14 we complete the proof of Theorems 1.2 and 1.3, giving general conditions on a model of random graph that are sufficient to imply the Alon conjecture. In Chapter 15 we make some closing remarks.

We mention that the reader interested only in the Alon conjecture for only $\mathcal{G}_{n,d}$ (i.e., the first part of Theorem 1.1) need not read Chapters 2, 4 (assuming a willing to believe Lemma 4.10), and 14 and Sections 3.7, 3.8, 5.4, and 6.4. Chapter 2 explains the problems with the trace method encountered in [**Fri91**]. Sections 3.7 and 3.8 and Chapter 4 concern themselves with the second part of Theorem 1.1 (the close to matching bound on how many graphs fail the $2\sqrt{d-1} + \epsilon$ bound). Section 5.4 explains the new aspects in our approach to the Alon conjecture; this subsection is not essential to the exposition (but probably is helpful). Section 6.4 and Chapter 14 involve the Alon conjecture for $\mathcal{H}_{n,d}, \mathcal{I}_{n,d}, \mathcal{J}_{n,d}$.

Throughout the rest of this paper we will work with $\mathcal{G}_{n,d}$ unless we explicitly say otherwise, and we understand d to be a fixed integer at least 3. At times we insist that d be even (for example, in dealing with $\mathcal{G}_{n,d}$ and $\mathcal{H}_{n,d}$).

CHAPTER 2

Problems with the Standard Trace Method

In this section we summarize the trace method used in [**Fri91**], and why this method cannot prove Alon's conjecture. During this section we will review some of the ideas of [**Fri91**], involving asymptotic expansions of various types of traces, which we modify in later sections to complete our proof of Alon's conjecture.

1. The Trace Method

We begin by recalling the trace method as used in [**Fri91**], and why it did not yield the Alon conjecture.

The trace method (see [**Wig55, Gem80, FK81, McK81, BS87, Fri91**], for example) determines information on the eigenvalues of a random graph in a certain probability space by computing the expected value of a sufficiently high power of the adjacency matrix, A; this expected value equals the expected value of the sum of that power of the eigenvalues, since

$$\text{Trace}\left(A^k\right) = \lambda_1^k + \cdots + \lambda_n^k.$$

Now Trace $\left(A^k\right)$ may also be interpreted as the number of closed walks (i.e., walks (see Chapter 3.1) in the graph that start and end at the same vertex) of length k. Now restrict our discussion to $\mathcal{G}_{n,d}$. A word, $w = \sigma_1 \ldots \sigma_k$, of length k over the alphabet

$$\Pi = \{\pi_1, \pi_1^{-1}, \ldots, \pi_{d/2}, \pi_{d/2}^{-1}\}$$

(i.e. each $\sigma_i \in \Pi$), determines a random permutation, and the i,j-th entry of A^k, is just the number of words, w, of length k, taking i to j. But given a word, w, the probability, $P(w)$, that w takes i to i is clearly independent of i. Hence we have

$$\mathrm{E}\left[\text{Trace}\left(A^k\right)\right] = n \sum_{w \in \Pi^k} P(w)$$

In [**BS87**], Broder and Shamir estimated the right-hand-side of the above equality to obtain an estimate on λ_2. This analysis was refined in [**Fri91**]. We review the ideas there.

First, a word, w, is said to be *irreducible* if w contains no consecutive occurrence of σ, σ^{-1}. It is well-known that any word, w, has a unique *reduction* to an irreducible word[1] (or reduced word), w', obtained from w by repeatedly discarding any consecutive occurrences of σ and σ^{-1} in w, and $P(w) = P(w')$. Similarly a

---

[1]In fact, the irreducible word has length which is its distance to the identity in the Cayley graph over the free group on $d/2$ elements (see [**FTP83**], Sections 1 and 7 of chapter 1). Alternatively, see Proposition 2.5 of [**DD89**] or Theorem 1 of [**Joh90**] (this theorem says that a free group on a set, X, is in one-to-one correspondence with the set of reduced words, $X \cup X^{-1}$, which means that every word over $X \cup X^{-1}$ has a unique corresponding reduced word; here "reduced" is our "irreducible").

walk is said to be *irreducible* if it contains no occurrence of a step along an edge immediately followed by the reverse step along that edge[2]. Similarly, every irreducible walk has a unique reduction. Let Irred_k be the set of irreducible words of length k, and let $\text{IrredTr}(A, k)$ be the number of irreducible closed walks of length k in G[3]. We shall see that to evaluate the expected value of $\text{Trace}(A^m)$ it suffices, in a sense (namely that of equation (3) below), to evaluate
$$\mathrm{E}\left[\text{IrredTr}(A,k)\right] = n \sum_{w \in \text{Irred}_k} P(w),$$
for $k = m, m-2, \ldots$. It is easy to see that for any fixed word, w, we have a power series expansion
$$P(w) = P_0(w) + \frac{P_1(w)}{n} + \frac{P_2(w)}{n^2} + \cdots$$
(see, for example, Theorem 5.5).

As examples, we note that for a random permutation, π, on $\{1, \ldots, n\}$, the probability that the sequence $1, \pi(1), \pi^2(1), \ldots$ first returns to 1 at $\pi^k(1)$ (i.e., the probability that 1 lies on a cycle of length exactly k) is $1/n$ for $k = 1, \ldots, n$. It follows that $P(\pi_1^m) = \phi(m)/n$, where $\phi(m)$ is the number of positive integral divisors of m, assuming $m \leq n$. In this example $P_1(w) = \phi(m)$ involves number theoretic properties of m. For a second example, we first remark that π^m maps a fixed vertex to a different vertex with probability $1 - \bigl(\phi(m)/n\bigr)$, and to each of the $n-1$ different vertices with the same probability. It is then easy to see that
$$P(\pi_1^{m_1}\pi_2^{m_2}) = \frac{\phi(m_1)\phi(m_2)}{n^2} + \frac{\bigl(n-\phi(m_1)\bigr)\bigl(n-\phi(m_2)\bigr)}{n^2(n-1)}$$
provided that m_1, m_2 are at most n. If m_1, m_2 are at least 2, then the P_i are non-zero for $i \geq 1$ and involve number theoretic functions of m_1, m_2 for $i \geq 2$.

So set
$$(1) \qquad g_i(k) = \sum_{w \in \text{Irred}_k} P_{i+1}(w)$$
(we easily see $P_0(w) = 0$ for $w \in \text{Irred}_k$ and $k \geq 1$ and so $g_{-1}(k) = 0$ for $k \geq 1$).

DEFINITION 2.1. *A function, $f(k)$, on positive integers, k, is said to be d-Ramanujan if there is a polynomial $p = p(k)$ and a constant $c > 0$ such that*
$$|f(k) - (d-1)^k p(k)| \leq ck^c(d-1)^{k/2}$$
for all k. We call $(d-1)^k p(k)$ the principal term *of f, and $f(k) - (d-1)^k p(k)$ the* error term *(both terms are uniquely determined if $d > 2$).*

In [**Fri91**] it was shown (among other things) that for all $i \leq \sqrt{d-1}/2$ we have that g_i as above is d-Ramanujan. This, it turns out, gives a second eigenvalue bound of roughly $2\sqrt{d-1} + 2\log d + C + O(\log \log n / \log n)$ for a universal constant, C. We now explain why.

A standard counting and expansion argument is given in [**Fri91**] (specifically Theorem 3.1 there) to establish the following lemma.

---

[2] In the case of an edge that is a half-loop (see Chapter 3), a half-loop may not be traversed twice consecutively in an irreducible walk.

[3] We have admittedly defined $\text{IrredTr}(A, k)$ in terms of G, but we shall soon see (Lemma 2.3) that $\text{IrredTr}(A, k)$ can be defined as a polynomial in A and d.

LEMMA 2.2. *For fixed even $d \geq 4$ there is an $\eta > 0$ such that with probability $1 - n^{1-d} + O(n^{2-2d})$ we have that a G in $\mathcal{G}_{n,d}$ has $\max(\lambda_2, -\lambda_n) \leq d - \eta$ (also with probability $n^{1-d} + O(n^{2-2d})$ we have that $\lambda_2 = d$).*

Next to $\lambda_1 = d$, one (or both) of λ_2, λ_n is the next largest eigenvalue in absolute value; Lemma 2.2, by bounding the eigenvalues other than λ_1, will eventually be used to show that the g_i of equation (1) are essentially determined, for small i, by λ_1's "contribution" to $\mathrm{IrredTr}\,(A, k)$ (see below).

Next we establish the precise relationship between the traces of the A^k and the $\mathrm{IrredTr}\,(A, k)$. Let A_k be the matrix whose i, j-th entry is the number of irreducible walks of length k from i to j.

LEMMA 2.3. *The A_k are given by*
$$A_k = q_k(A),$$
where q_k is the degree k polynomial given via
$$(2) \qquad q_k(2\sqrt{d-1}\cos\theta) = \left(\sqrt{d-1}\right)^k \left(\frac{2}{d-1}\cos k\theta + \frac{d-2}{d-1}\frac{\sin(k+1)\theta}{\sin\theta}\right)$$
(which is a type of Chebyshev polynomial); alternatively we have $q_1(x) = x$, $q_2(x) = x^2 - d$, and for $k \geq 3$ we have
$$q_k(x) = x\, q_{k-1}(x) - (d-1)q_{k-2}(x).$$
Also
$$\mathrm{IrredTr}\,(A, k) = \mathrm{Trace}\,(A_k) = \sum_{i=1}^{n} q_k(\lambda_i).$$

The proof is given in [**LPS86**] and [**Fri91**] (specifically Lemma 3.3, page 356, in [**Fri91**]; the F_k's there are the A_k's here). In Chapter 10 we shall use the fact that for fixed λ, $q_k = q_k(\lambda)$ satisfy the recurrence
$$\left(\sigma_k^2 - \lambda\sigma_k + (d-1)\right)q_k = 0,$$
where σ_k is the "shift in k" operator (i.e., $\sigma_k q_k = q_{k+1}$)

To go the other way we note:
$$(3) \qquad A^k = \sum_{i=k,k-2,k-4,\ldots} N_{k,i} A_i,$$
where $N_{k,i}$ is the number of words of length k that reduce to a given irreducible word of length i. Thus
$$\mathrm{Trace}\,(A^k) = \sum_{i=k,k-2,k-4,\ldots} N_{k,i}\, \mathrm{IrredTr}\,(A, i).$$

LEMMA 2.4. *For k, i even we have*
$$N_{k,i} \leq \left(2\sqrt{d-1}\right)^k (d-1)^{-i/2}\sqrt{(d-1)/d}$$
if $i > 0$ and
$$N_{k,0} \leq \left(2\sqrt{d-1}\right)^k.$$

An exact formula for $N_{k,i}$ is given in [**McK81**]. A weaker estimate than the above lemma was used in [**Fri91**]. The proof of this estimate is a simple spectral argument used by Buck (see [**Buc86, Fri03**]).

PROOF. Consider the adjacency matrix, A_T, of the infinite d-regular tree, T. Our proof requires the following sublemma.

SUBLEMMA 2.5. *A_T has norm $\leq 2\sqrt{d-1}$.*

Actually, it is well-known that the norm of A_T is exactly $2\sqrt{d-1}$ (see, for example, page 9 of [**Woe00**], and the theorems on λ_1 in Chapter 3 here). However, the proof below is simple and generalizes to many other situations (and can be used in many cases to determine the exact norm of A_T).

PROOF. (**of Sublemma 2.5**) Let f be a function in $L^2(T)$, i.e., a function on the vertices of T whose sum of squares of values is finite. Fix a vertex, v_0, of T, to be viewed as the root of T; the *children* of a vertex, v, are defined to be those vertices adjacent to v and of greater distance than v is to v_0. We have

$$(A_T f, f) = \sum_v \sum_{w \in \text{children}(v)} 2f(v)f(w)$$

which, by Cauchy-Schwarz, is

$$\leq \sum_v \sum_{w \in \text{children}(v)} \left(f^2(w)\sqrt{d-1} + \frac{f^2(v)}{\sqrt{d-1}} \right)$$

$$= f^2(v_0) d \Big/ \sqrt{d-1} + \sum_{v \neq v_0} f^2(v) 2\sqrt{d-1}.$$

$$\leq \sum_v f^2(v) 2\sqrt{d-1} = 2\sqrt{d-1} \|f\|^2.$$

Thus the norm of A_T is $\leq 2\sqrt{d-1}$. □

(To see that the norm of A_T is exactly $2\sqrt{d-1}$ we find functions, f, (of finite support) for which the applications of Cauchy-Schwarz in the above proof are "usually" tight. Namely, we can take $f(v) = (d-1)^{-\text{dist}(v,v_0)/2}$ for $\text{dist}(v, v_0) \leq s$ and $f(v) = 0$ otherwise, where s is a parameter which tends to ∞. This technique works for some other graphs.)

We resume the proof of Lemma 2.4. Let v be a vertex of T, and let S be the vertices of distance i to s. Then $|S| N_{k,i}$ is the dot product of $A_T^k \chi_{\{v\}}$ with χ_S, where χ_U denotes the characteristic function of U, i.e. the function that is 1 on U and 0 elsewhere. So by Cauchy-Schwarz

$$|S| N_{k,i} = (A_T^k \chi_{\{v\}}, \chi_S) \leq \|A_T\|^k |\chi_{\{v\}}| \, |\chi_S| = \left(2\sqrt{d-1}\right)^k \sqrt{|S|}.$$

We finish with the fact that $|S| = 1$ if $i = 0$, and otherwise $|S| = d(d-1)^{i-1}$. □

Notice that clearly $N_{k,k} = 1$, and so for $i = k$ Lemma 2.4 is off by a multiplicative factor of roughly 2^k; according to [**McK81, FTP83**], the Lemma 2.4 estimate of $N_{k,0}$ is off by roughly a factor of $k^{3/2}$. The roughness of Lemma 2.4 is unimportant for our purposes.

Now notice that by Lemmas 2.2 and 2.3 we have

$$\mathrm{E}\left[\text{IrredTr}(A, k)\right] = q_k(d)\left(1 + n^{1-d} + O(n^{2-2d})\right) + \text{error},$$

where

$$|\text{error}| \leq (n-1) \max_{|\lambda| \leq d-\eta} |q_k(\lambda)|.$$

It is easy to see (see [**Fri91**]) that $q_k(d) = (d-1)^k$, and for some $\alpha > 0$ we have
$$\max_{|\lambda| \le d-\eta} |q_k(\lambda)| \le (d-1-\alpha)^k ck,$$
for an absolute constant c (with any η as in Lemma 2.2). We wish to draw some conclusions about the principal term of the g_i's. We need the following lemma:

LEMMA 2.6. *For fixed d, r there is a constant, c, such that for $k \ge 1$ we have that in $\mathcal{G}_{n,d}$*
$$E[\mathrm{IrredTr}\,(A, k)] = g_0(k) + \frac{g_1(k)}{n} + \frac{g_2(k)}{n^2} + \cdots + \frac{g_{r-1}(k)}{n^{r-1}} + \mathrm{error},$$
where
$$|\mathrm{error}| \le c(d-1)^{k-1} k^{4r+2}/n^r.$$

PROOF. This follows from the calculations on page 352 of [**Fri91**]; for each i, the f_i in [**Fri91**] is the polynomial in the principal term of g_i (and we mean f_i corresponds precisely to g_i, not g_{i-1} or g_{i+1}). (Actually we shall later[4] see that the $4r + 2$ in the error term estimate can be replaced by $4r$.) □

We now take k of order $\log^2 n$ and use standard facts about expansion and expansion's control on eigenvalues (namely our Lemma 2.2) to conclude, as in [**Fri91**], the following theorem.

THEOREM 2.7. *Let $g_0, g_1, \ldots, g_r$ be d-Ramanujan for some $r \le d$. Then the principal term of g_i vanishes for $1 \le i \le r$, and the principal term of g_0 is $d(d-1)^{k-1}$.*

PROOF. See Theorem 3.5 of [**Fri91**]. □

We next apply Lemma 2.4 to estimate the expected value of the trace of A^k where k is roughly
$$(4) \qquad h(n, r, d) = \frac{(r+1)\log n}{\log\bigl(d/(2\sqrt{d-1})\bigr)},$$
as in [**Fri91**], in order to obtain the following theorem.

THEOREM 2.8. *With the same hypotheses as Theorem 2.7, we have (in $\mathcal{G}_{n,d}$)*
$$(5) \qquad E\left[\sum_{i=2}^n \lambda_i^k\right] \le \rho^k$$
for all $k \le h(n, r, d)$, with h as above, where
$$(6) \qquad \rho = 2\sqrt{d-1}\,\bigl(d/(2\sqrt{d-1})\bigr)^{1/(r+1)} \left(1 + \frac{c \log\log n}{\log n}\right)$$
and c depends only on r, d.

PROOF. First we take $k = \lfloor h(n, r, d) \rfloor$ and find that ρ can be taken as above. For smaller k we appeal to Jensen's inequality. See [**Fri91**] for details. □

---

[4] This stems from the fact that in [**Fri91**], the $e^{(r+1)k/n} k^{2r+2}$ just above equation (21) (page 352) could have been replaced with $e^{rk/n} k^{2r}$.

From Theorem 2.18 of [**Fri91**] we see that we can take r as large as $\lfloor \sqrt{d-1}/2 \rfloor$. With this value of r, using equation (6), we conclude that equation (5) holds with $\rho = 2\sqrt{d-1} + 2\log d + C + o(1)$ for an absolute constant, C, where $o(1)$ is a quantity that for fixed d tends to 0 (proportional to $\log\log n/\log n$) as $n \to \infty$.

Whenever equation (5) holds, then the expected value of $\max(|\lambda_2|, |\lambda_n|)$ is bounded by ρ. The Alon conjecture would be implied if one could obtain $\rho \leq 2\sqrt{d-1} + \epsilon$ for any $\epsilon > 0$.

2. Limitations of the Trace Expansion

In this subsection we will show that for some $i \leq O(\sqrt{d}\log d)$, g_i is not d-Ramanujan. We similarly show the part of Theorem 1.1 involving s. Both these facts are due to the possible occurrence of what we call *tangles*. Tangles and avoiding them are the main themes in this paper.

We begin by describing an example of a *tangle*, and its effect on eigenvalues and traces. Consider $\mathcal{G}_{n,d}$ for a fixed, even d and a variable n which we view as large. Fix an integer m with $1 \leq m \leq d/2$ (assume $d \geq 4$). Consider the event, $\mathcal{T}$, that
$$\pi_i(1) = 1 \text{ for } i = 1, \ldots m.$$
Clearly $\mathcal{T}$ occurs with probability $1/n^m$.

Assume $\mathcal{T}$ occurs in a fixed d-regular graph, $G = (V, E)$. Let W be the set of vertices of distance at least 2 to the vertex 1; W is a random set of vertices, but always of size at least $n - d - 1$. Consider the characteristic functions $\chi_{\{1\}}, \chi_W$, where χ_U is the function that is 1 on the vertices in U, and 0 elsewhere. Let
$$\mathcal{R}_A(v) = \frac{(Av, v)}{(v, v)}$$
be the Rayleigh quotient associated to the adjacency matrix, A, of G. The following lemma is well-known.

LEMMA 2.9. *Let A be a real, symmetric matrix. Let u, v be nonzero vectors with v orthogonal to u and Au. Then*
$$\lambda_2 \geq \min(\mathcal{R}_A(u), \mathcal{R}_A(v)).$$

PROOF. Let μ denote the min on the right-hand-side of the above inequality. By the hypothesis of the lemma, $(Au, v) = 0$; along with the symmetry of A, we have $(Av, u) = (v, Au) = 0$. If $w = \alpha u + \beta v$ with α, β scalars, we have
$$(Aw, w) = (Au, u)\alpha^2 + (Av, v)\beta^2 \geq \mu(u, u)\alpha^2 + \mu(v, v)\beta^2 = \mu(w, w).$$
It follows that the Rayleigh quotient of any vector in the span of u and v is at least μ. Since this span is a two-dimensional subspace, the max-min principle implies that $\lambda_2 \geq \mu$. □

We intend apply the above lemma with $u = \chi_{\{1\}}$ and $v = \chi_W$.
$$\mathcal{R}_A(\chi_{\{1\}}) \geq 2m$$
(since $(A\chi_U, \chi_U)$ counts twice the number of edges with both endpoints in U). Also
$$(A\chi_W, \chi_W) = (A\chi_V, \chi_V) - 2(A\chi_V, \chi_{V\setminus W}) + (A\chi_{V\setminus W}, \chi_{V\setminus W})$$
$$\geq (A\chi_V, \chi_V) - 2(A\chi_V, \chi_{V\setminus W}) \geq dn - 2d(d+1)$$

so
$$\mathcal{R}_A(\chi_W) \geq \frac{dn - 2d(d+1)}{n-d-1} = d - O(1/n)$$
viewing m, d as fixed. Since $\chi_{\{1\}}$ and $A\chi_{\{1\}}$ are supported in the neighbourhood of distance at most 1 from the vertex 1, χ_W is orthogonal to both of them. Lemma 2.9 now implies
$$\lambda_2 \geq \min(2m, d - O(1/n)).$$
Next consider the probability that
$$\pi_i(r) = r \text{ for } i = 1, \ldots, m$$
for at least one value of r. Inclusion/exclusion shows that the probability of this is at least
$$\sum_r \mathrm{Prob}\left\{\pi_i(r) = r \text{ for } i = 1, \ldots m\right\} -$$
$$\sum_{r,s} \mathrm{Prob}\left\{\pi_i(r) = r \text{ and } \pi_i(s) = s \text{ for } i = 1, \ldots m\right\}$$
$$\geq n^{1-m} - \binom{n}{2} n^{-2m}.$$
We summarize the above observations.

THEOREM 2.10. *For fixed integer m with $1 \leq m \leq d/2$, we have that $\lambda_2 \geq 2m$ for sufficiently large n with probability at least $n^{1-m} - (1/2)n^{2-2m}$.*

The proof of the above theorem did not exploit the fact that aside from having m self-loops, the vertex 1 is still adjacent to $d - 2m$ other vertices of a d-regular graph. We seek a stronger theorem that exploits this fact.

THEOREM 2.11. *For fixed integers $m \geq 1$ and $d \geq 4$, with $2m - 1 > \sqrt{d-1}$ and $m \leq d/2$, we have that $\lambda_2 > 2\sqrt{d-1}$ for sufficiently large n with probability at least $n^{1-m} - (1/2)n^{2-2m}$.*

(Notice that Theorem 2.10 would require $m > \sqrt{d-1}$ for the same conclusion.) We are very interested to know if one can prove Theorem 2.11 when $2m-1 = \sqrt{d-1}$ for integer m and even integer d. We expect not. (The situation where $2m - 1 = \sqrt{d-1}$ gives rise to what we will call a "critical tangle," and $2m - 1 \geq \sqrt{d-1}$ to a "supercritical tangle," in Chapter 4.)

PROOF. Note: in Theorems 3.13 and 4.2 we give a proof of a generalization of this theorem requiring far less calculation (but requiring more machinery).

Again, assume that $\pi_i(v_0) = v_0$ for $i = 1, \ldots, m$ and some v_0. It suffices to show $\lambda_2 > 2\sqrt{d-1}$ for sufficiently large n, under the assumption that $2m - 1 > \sqrt{d-1}$.

Let
$$\alpha(m) = (2m-1) + \frac{d-1}{2m-1}. \tag{7}$$

By Cauchy-Schwarz we have $\alpha > 2\sqrt{d-1}$ (equality does not hold, because $2m-1 \neq (d-1)/(2m-1)$ since $2m - 1 > \sqrt{d-1}$).

Let $\rho(v)$ denote v's distance to v_0. For a fixed r, let
$$f(v) = \begin{cases} (2m-1)^{-\rho} & \text{if } \rho \leq r, \\ 0 & \text{otherwise,} \end{cases}$$

where $\rho = \rho(v)$. It is easy to check that $(Af)(v) \geq \alpha f(v)$ provided that $\rho(v) < r$ (this includes the case $v = v_0$, since $\rho(v_0) = 0$, but checking $v = v_0$ is a bit different from the other cases). It follows that

$$(8) \qquad \frac{(Af,f)}{(f,f)} \geq \frac{\alpha(f,f)_{r-1}}{(f,f)_r},$$

where

$$(f,f)_t = \sum_{\rho(v) \leq t} f^2(v).$$

But $1 = f^2(v_0) \leq (f,f)_{r-1}$ if $r \geq 1$, and also

$$(f,f)_r \leq (f,f)_{r-1} + (d-2m)(d-1)^{r-1}(2m-1)^{-2r}$$

(since, by induction, the number of vertices at distance r from v is at most $(d-2m)(d-1)^{r-1}$.) So

$$(9) \qquad \frac{(f,f)_r}{(f,f)_{r-1}}$$

can be made arbitrarily close to 1 by taking r sufficiently large (since $(d-1)(2m-1)^{-2} < 1$).

Let $\mathcal{R}$ be the Rayleigh quotient of A. The last paragraph, especially equations (8) and (9) implies that for $\alpha' < \alpha$ there is an $r = r(\alpha')$ such that $\mathcal{R}(f) \geq \alpha'$.

So let N be those vertices of distance 1 or 0 to the support of f; the size, $|N|$, of N is bounded as a function of d and r. The function f is orthogonal to $g = \chi_{V \setminus N}$ and Ag, and counting edges as before we see

$$\mathcal{R}(g) \geq \frac{d|V| - 2d|N|}{|V| - |N|} = d - O(|N|d/|V|).$$

It follows that by taking n sufficiently large, we can make $\lambda_2 \geq \alpha'$; since $\alpha > 2\sqrt{d-1}$, we can choose $\alpha' > 2\sqrt{d-1}$, making $\lambda_2 > 2\sqrt{d-1}$. □

Theorem 2.11 proves the part of Theorem 1.1 involving s, by taking $s = m-1$ with m as small as possible (namely $m = \lfloor (\sqrt{d-1} + 1)/2 \rfloor + 1$). The analogous parts of Theorems 1.2 and 1.3 are slightly trickier, since the "tangle" involved has automorphisms; we shall delay their proof (see Theorem 4.13) until we give a more involved discussion of tangles in Chapter 4.

Notice that our proof is really computing the norm of A_H where H is the d-regular graph with the vertex 1 having m self-loops, and which is a (an infinite) tree when these loops are removed. The function f as above shows that A_H's norm is at least α. The statement and proof of Sublemma 2.5 for the d-regular tree applies to the above tree (with $2m-1$ replacing $\sqrt{d-1}$, and with α replacing $2\sqrt{d-1}$). In this way our proof of Theorem 2.11 is very much like one proof of the Alon-Boppana theorem (see [**Nil91, Fri93**]).

The discussion in this section leads to the following theorem.

THEOREM 2.12. *There is an absolute constant (independent of d), C, such that the g_i of equation (1) cannot be d-Ramanujan for all $i \leq C\sqrt{d} \log d$.*

PROOF. We fix an integer s to be chosen later with

$$\left(\sqrt{d-1} - 1\right)/2 < s < d/2.$$

Set $s+1 = m$ and apply Theorem 2.11. Since $\alpha \geq 2s+1$ with α as in equation (7), for k even we have that $\lambda_2 \geq 2s$ with probability at least $n^{-s} + O(n^{-s-1})$. Thus
$$\mathrm{E}\left[\lambda_2^k\right]^{1/k} \geq \left(n^{-s} + O(n^{-s-1})\right)^{1/k} 2s.$$
According to Theorem 2.8, if $g_0, \ldots, g_r$ are d-Ramanujan for some $r \leq d$, then we have
$$\mathrm{E}\left[\lambda_2^k\right]^{1/k} \leq \rho,$$
with ρ as in equation (6), provided that k is even and bounded by $h(n, r, d)$ as in Theorem 2.8. For some constant c we have that for any C and for $r = C\sqrt{d}\log d$, equation (6) gives
$$\rho = 2\sqrt{d-1}\left(1 + cC^{-1}d^{-1/2} + c(\log n)^{-1}\log \log n\right).$$
In other words,

(10) $\quad \left(n^{-s} + O(n^{-s-1})\right)^{1/k} 2s \leq 2\sqrt{d-1}\left(1 + cC^{-1}d^{-1/2} + c(\log n)^{-1}\log\log n\right).$

Take k even and as close to $h(n, r, d)$ as possible; note that by equation (4),
$$\frac{\log n}{h} = \frac{\log\left(d/(2\sqrt{d-1})\right)}{r+1} \leq \frac{\log d}{(C\sqrt{d}\log d)+1} \leq 1/\left(C\sqrt{d}\right);$$
hence, taking $n \to \infty$ in equation (10) implies that for a universal constant, c', we have
$$e^{-sd^{-1/2}/C} 2s \leq 2\sqrt{d-1}\left(1 + cC^{-1}d^{-1/2}\right) \leq 2\sqrt{d-1}\left(1 + cC^{-1}\right).$$
Choosing $s = C\sqrt{d}$ and dividing by 2 yields
$$C\sqrt{d}/e \leq \sqrt{d-1}\left(1 + cC^{-1}\right).$$
Choosing C large enough so that $C/e > 1 + (c/C)$ makes this impossible. $\square$

We have proven that not all g_i are d-Ramanujan for $i \leq r$ where $r = C\sqrt{d}\log d$. Notice that in our terminology, Theorem 2.18 of [**Fri91**] says that g_i is d-Ramanujan for $i \leq \lfloor\sqrt{d-1}/2\rfloor - 1$; again, for each i the f_i in [**Fri91**] is the polynomial in the principal part of our g_i. This leaves the question of whether Theorem 2.12 can be improved to an r value closer to $\lfloor\sqrt{d-1}/2\rfloor - 1$; we conjecture that it can be improved to $r = \lfloor(\sqrt{d-1}+1)/2\rfloor$, and that the tangle with $m = \lfloor(\sqrt{d-1}+1)/2\rfloor + 1$ already "causes" this g_r (or a lower one) to fail to be d-Ramanujan.

This also leaves open the question of what can be said about the g_i that are not d-Ramanujan. Perhaps such $g_i = g_i(k)$ are a sum of $\nu^k p_\nu(k)$ over various ν with some added error term. In this paper we avoid this issue, working with a modified trace (i.e., "selective" traces) for which the corresponding g_i are d-Ramanujan.

CHAPTER 3

Background and Terminology

In this section we review some ideas and techniques from the literature needed here. We also give some convenient terminology that is not completely standard.

1. Graph Terminology

We use some nonstandard notions in graph theory, and we carefully explain all our terminology and notions here.

A *directed graph*, G, consists of a set of vertices, $V = V_G$, a set of edges, $E = E_G$, and an incidence map, $i = i_G \colon E \to V \times V$; if $i(e) = (u,v)$ we will write $e \sim (u,v)$, say that e is of *type* (u,v), and say that e originates in u and terminates in v. (If i is injective then it is usually safe to view E as a subset of $V \times V$, and we say that G has no multiple edges.) The adjacency matrix, $A = A_G$, is a square matrix indexed on V, where $A(u,v)$ counts the number of edges of type (u,v). The outdegree at $v \in V$ is the row sum of A at v, i.e., the number of edges originating in v; the indegree is the column sum or number of edges terminating in v.

A *graph*, G, is a directed graph, $\widehat{G}$, such that each edge of type (u,v) is "paired" with an "opposite edge" of type (v,u); in other words, we have a map $\mathrm{opp} = \mathrm{opp}_G \colon E_{\widehat{G}} \to E_{\widehat{G}}$, such that $\mathrm{opp}(\mathrm{opp})$ is the identity, and if $e \in E_{\widehat{G}}$ has $e \sim (u,v)$, then $\mathrm{opp}(e) \sim (v,u)$; in other words, the edges $E_{\widehat{G}}$ come in "pairs," except that a self-loop, i.e., an $e \in E_{\widehat{G}}$ with $e \sim (v,v)$, can be paired with itself (which is a "half-loop" in the terminology of [**Fri93**]) or paired with another self-loop at v (which is a "whole-loop"). Half-loops about v contribute 1 to the adjacency matrix entry at v,v (i.e., contribute 1 to $A(v,v)$), and whole-loops contribute 2. In this paper we primarily work with whole-loops, needing half-loops only in the model $\mathcal{J}_{n,d}$. We refer to the *(undirected) edges*, E_G, of a graph, G, as the set of "pairs" of edges, $\{e, \mathrm{opp}(e)\}$. G's vertex set and adjacency matrix are just those of the directed graph, $\widehat{G}$, i.e., $V_G = V_{\widehat{G}}$ and $A_G = A_{\widehat{G}}$.

A *numbering* of a set, S, is a bijection $\iota \colon S \to \{1, 2, \ldots, s\}$, where $s = |S|$. A *partial numbering* of a set, S, is a numbering of some subset, S', of S (we allow S' to be empty, in which case none of S is numbered). We can speak of a graph, directed or not, as having numbered or partially numbered vertices and/or edges. A numbering can be viewed as a total ordering.

Each letter $\pi \in \Pi = \{\pi_1, \pi_1^{-1}, \ldots, \pi_{d/2}^{-1}\}$ has its associated inverse, $\pi^{-1} \in \Pi$, and every word $w = \sigma_1 \ldots \sigma_k$ over Π has its associated inverse, $w^{-1} = \sigma_k^{-1} \ldots \sigma_1^{-1}$. If $\mathcal{W}$ is any set of words over Π, then a $\mathcal{W}$-*labelling* of an undirected graph, G, is a map or "labelling" $\mathcal{L} \colon E_{\widehat{G}} \to \mathcal{W}$ such that $\mathcal{L}(\mathrm{opp}(e)) = (\mathcal{L}(e))^{-1}$ for each $e \in E_{\widehat{G}}$. For example, any graph $G \in \mathcal{G}_{n,d}$ automatically comes with a Π-labelling, namely $(i, \pi_j(i))$ is labelled π_j, and $(i, \pi_j^{-1}(i))$ is labelled π_j^{-1}.

An *orientation* of an undirected graph, G, is the distinguishing for each $e \in E_G$ of one of the two directed edges corresponding to e.

The following definition is special to this paper.

DEFINITION 3.1. *Fix sets V, E and a set of words, $\mathcal{W}$, over Π, with $\mathcal{W}^{-1} = \mathcal{W}$. A* structural map *is a map $s\colon E \to \mathcal{W} \times V \times V$. A structural map defines a unique $\mathcal{W}$-labelled, oriented graph, G, with $V_G = V$ and $E_G = E$, as follows: for each $e \in E$ with $s(e) = (\sigma, u, v)$, we form a directed edge of type (u, v) labelled σ and declare it distinguished, and pair it with a directed edge of type (v, u) labelled σ^{-1}.*

2. Variable-Length Graphs and Subdivisions

In this paper we will work with graphs that have large or infinite parts of them being paths or regular trees. In this case we can easily eliminate all the vertices in these parts by working with "variable-length graphs." This leads to simpler calculations (in Theorem 3.6, that is crucial to Lemma 9.2, and in Theorem 6.6). This also gives us a notion of regular tree of non-integral degree, in Theorem 3.13.

Recall (see [**SW49, AFKM86, HMS91, Fri93**]) that a *VLG* or *variable-length graph* (respectively, *directed VLG* or *directed variable-length graph*) is a graph (respectively, directed graph) with an assignment of a positive integer to each edge called the edge's *length*[1]. The length of a walk in a VLG is the sum of the lengths of its edges (each length is counted the number of times the edge appears in the walk).

A graph can be regarded as a VLG with all edge lengths 1. A VLG whose edge lengths are all 1 can be identified with its underlying graph.

A *bead* in a directed graph (respectively graph) is a vertex with indegree and outdegree 1 (respectively, degree 2) and without a self-loop. A *beaded path* is a path where every vertex except possibly the endpoints are beads.

DEFINITION 3.2. *Let G be a directed VLG. To* subdivide *an edge, e, from u to v and of length ℓ, in G is to replace e with a beaded path of length ℓ from u to v in G (introducing $\ell - 1$ new vertices). A* subdivided form *of G is a graph, G_{sbd}, obtained by subdividing all edges of G. The same definition is made for VLG's, omitting the word "directed" everywhere, provided no half-loops are of length 2 or greater*[2].

It should be clear that countings walks of certain types in a directed VLG, G, should translate to an appropriate similar counting in G_{sbd}, and vice versa. We next define an opposite of subdivision, suppression, and a vast generalization of suppression, realization.

DEFINITION 3.3. *Let G be a strongly connected directed graph, with $W \subset V_G$. The* realization *of G to with vertex set W denoted $G|_W$, is the directed VLG on vertices W with the following set of edges. We create $G|_W$ one edge from u to v (for $u, v \in W$) of length k for each walk from u to v in G of length k that contains no W vertices except as the first and last vertices. (So self-loops or edges in G

---

[1]In Shannon's terminology of [**SW49**], Chapter 1, Chapter 1, the edges have various "times" (as opposed to "lengths") such as a dot versus a dash in Morse code.

[2]This restriction will be explained just before Proposition 3.7. Actually, we can define a notion of subdivision for all half-loops of odd length, but in this paper we use half-loops only of length 1.

involving W vertices appear in $G|_W$, since we regard self-loops or edges as walks with no vertices except the first and last vertices.)

The notion of realization appears constrained coding theory (called "fusion" in [**HMS91**], for example, and not given a name in [**AFKM86**]; see also [**Fri93**]). We remark that if $V_G \setminus W$ contains a cycle, then $G|_W$ has infinitely many edges.

DEFINITION 3.4. *Let G be a strongly connected directed graph, with $U \subset V_G$ a subset of beads in G such that U contains no cycle. The* suppression *of U (in G), denoted $G\,[U]_{\mathrm{sup}}$, is the realization of G with vertex set $V_G \setminus U$.*

The subdivision (by the suppressed vertices) of a suppression returns the original directed graph.

We remark that if G is a Π-labelled graph, then any suppression in G has a natural Π^+-labelling, where Π^+ is the set of words on Π of length ≥ 1.

3. λ_1 of a VLG

Let G be a VLG (directed or undirected). For $u, v \in V_G$ and a non-negative integer, k, let $c_G(u, v; k)$ denote the number of walks of length k from u to v in G. We will use standard Perron-Frobenius theory (see Sections 1.3 and 7.1 in [**Kit98**] or Chapters 1 and 6 in [**Sen81**]), which includes the rest of this paragraph. Assume that G is strongly connected, i.e., for each $u, v \in V_G$ we have $c_G(u, v; k) > 0$ for some k. Let $d = d_G$ be the period of G, i.e., the greatest common divisor of the lengths of all closed walks in G. Then all limits

$$\limsup_{k \to \infty} \bigl(c_G(u, v; k)\bigr)^{1/k}, \quad \lim_{k \to \infty} \bigl(c_G(v, v; kd)\bigr)^{1/kd}$$

exist and are all equal (so independent of u, v); we define this common limit to be $\lambda_1(G)$, the *Perron value* of G. It is easy to see that $c_G(v, v; k) \leq \lambda_1^k$, using that $c_G(v, v; k_1) c_G(v, v; k_2) \geq c_G(v, v; k_1 + k_2)$. If G is a finite graph, then $\lambda_1(G)$ is just the usual Perron-Frobenius (largest) eigenvalue of A_G.

In directed graphs, suppression, realization, and subdivision preserve walk counts (i.e., the $c(u, v; k)$'s) between appropriate vertices (those present in the two graphs in question). Therefore these operations also preserve λ_1.

If G is not strongly connected, we can define $\lambda_1(G)$ to be the supremum of λ_1 of all the strongly connected components of G, or equivalently as the supremum over all $\bigl(c_G(v, v; k)\bigr)^{1/k}$.

One can equivalently define $\lambda_1(G)$ with $\tilde{c}_G$ replacing c_G, where $\tilde{c}_G(u, v; k)$ is the number of walks of length at most k. This is a sensible definition of $\lambda_1(G)$ when we allow non-integral edge lengths[3]. One can also extend all these definitions to graphs with positively weighted edges, where the weight of a walk becomes the product of its edge weights, and where c_G or $\tilde{c}_G$ sums the weights of the walks.

4. Shannon's Algorithm and Formal Series

Shannon gives the following algorithm (see [**SW49**], Chapter 1, Section 1) for computing $\lambda_1(G)$ (or the "valence" or "capacity") of a finite graph: let $Z_G = Z_G(z)$ be the matrix whose i, j entry is the sum of z^ℓ over all edge lengths, ℓ, of edges

---

[3]Why should a "dash" in Morse code be precisely an integral multiple of a "dot"?

from i to j, with z a formal parameter. Then $\lambda_1(G)$ is the reciprocal of the smallest real root in z of

(11) $$\det(I - Z_G(z)) = 0.$$

In this section we explain variants of this theorem/algorithm that hold for infinite VLG's. We first give some conventions that we will use with formal power series.

By a *non-negative power series* we mean a series $f(z) = \sum_{k=0}^\infty a_k z^k$, with a_k non-negative reals. We say that f is the *generating function* of the a_k. At times we view f as a formal power series, but we will also have cause to evaluate f at non-negative reals (as a possibly diverging infinite sum); it is easy to see that if $f(z_0)$ converges for a positive z_0, then f is continuous on $[0, z_0]$, and if $f(z_0)$ diverges then $f(y_0)$ gets arbitrarily large as y_0 approaches z_0.

For such an f, f's radius of convergence is

$$\rho(f) = \limsup_{k \to \infty} a_k^{1/k}.$$

The function f has a singularity at $z = \rho$, and $f(z_0) = \sum a_k z_0^k$ diverges for $z_0 > \rho$. If the singularity at $z = \rho$ is a pole (e.g., when $f(z)$ is a rational function), then $f(z) \to +\infty$ as $z \to \rho$ from the left.

N.B.: We do not identify a formal power series with any of its analytic extensions unless we specifically say so. For example, $f(z) = 1 + z + z^2 + \cdots$ has only the value $+\infty$ for real $z > 1$ unless we explicitly say to the contrary.

For a VLG, G, we set

$$M_G(z) = I + Z_G(z) + Z_G^2(z) + \cdots$$

We have

(12) $$\bigl(M_G(z)\bigr)_{u,v} = \sum_{k=0}^\infty c_G(u, v; k) z^k.$$

We say that a non-negative power series, $f(z) = \sum a_k z^k$, *majorizes* another one, $g(z) = \sum b_k z^k$, if

$$a_1 + \cdots + a_j \geq b_1 + \ldots + b_j$$

for all $j \geq 1$. If so, then $f(z_0) \geq g(z_0)$ for any $z_0 \in [0, 1]$ (with appropriate conventions on the value $+\infty$). Given VLG's, G and H, we say that G *majorizes* H if $Z_G(z)$ majorizes $Z_H(z)$ entry by entry; equivalently, there is an endpoint preserving injection from E_H to E_G that does not increase edge lengths. If so, clearly $M_G(z)$ majorizes $M_H(z)$ entry by entry.

THEOREM 3.5. *Let G be a strongly connected VLG, directed or undirected, on a countable number of vertices and edges. The following are equal:*

(1) $1/\lambda_1(G)$, *and*
(2) *the radius of convergence of any entry of $M_G(z)$.*

If G has a finite number of vertices, then the above two numbers also equal the following two:

(3) *the supremum of positive z such that $Z_G(z)$ converges (in each entry) and has largest (i.e., Perron-Frobenius) eigenvalue less than 1, and*
(4) *the supremum of positive z such that $Z_G(z)$ converges and $\det(I - Z_G(y)) > 0$ for all y with $0 < y < z$.*

If G has a finite number of vertices, and if the entries $Z_G(z)$ are all rational functions of z, then the four above quantities equal

(5) the first positive solution to $\det(I - Z_G(z)) = 0$.

Finally, all entries of $Z_G(z)$ will be rational functions of z whenever G is finite (i.e., has finitely many vertices and edges) or is a VLG realization of finite graph.

PROOF. (1)=(2): Clear from equation (12).

(2)=(3): Let B be a non-negative matrix. According to Perron-Frobenius theory, all the eigenvalues of B are of absolute value at most $\lambda_1(B)$, with equality only when the eigenvalue has the same algebraic and geometric multiplicity. It now follows from Jordan canonical form that a finite dimensional, non-negative matrix, B, has largest (Perron-Frobenius) eigenvalue less than 1 iff $I + B + B^2 + \cdots$ converges.

(3)=(4): If $\lambda_1(Z_G(z_0)) < 1$, then $\lambda_1(Z_G(y_0)) < 1$ for all $y_0 < z_0$, and hence $\det(I - Z_G(y_0)) > 0$ for such y_0. Consider the first z_0 for which $\lambda_1(Z_G(z_0)) < 1$ fails to hold (such a z_0 exists by the continuity of λ_1 as a function of its entries). Either $Z_G(z_0)$ does not converge, or else by continuity of λ_1 we have $\lambda_1(Z_G(z_0)) = 1$ and hence $\det(I - Z_G(z_0)) = 0$.

(4)=(5): By the above paragraph, if suffices to show that $Z_G(z)$ cannot have a pole (in any of its entries) at the first z_0 for which $\lambda_1(Z_G(z_0)) < 1$ fails to hold. But if $Z_G(z)$ has a pole in some entry at $z = z_0$, then for some $v \in V_G$ and positive integer, k, $Z_G^k(z)_{v,v}$ has a pole at $z = z_0$, by the strong connectivity of G. But $\lambda_1^k(Z_G(z)) \geq Z_G^k(z)_{v,v}$ for any z, and the latter tends to $+\infty$ as $z \to z_0$ from the left.

Last part of the theorem: Let G' be a realization of a finite VLG, G, on the set $U \subset V_G$. For $u_1, u_2 \in U$, we shall calculate the (u_1, u_2)-entry of $Z_{G'}(z)$. We claim this entry is the (u_1, u_2)-entry of $Z_G(z)$, plus the (u_1, u_2)-entry

$$v^{\mathrm{T}}(I + Z_{\bar{G}}(z) + Z_{\bar{G}}^2(z) + \cdots)w = v^{\mathrm{T}}(I - Z_{\bar{G}}(z))^{-1} w,$$

where $\bar{U}$ is the induced subgraph of G on the vertex set $\bar{U} = V_G \setminus U$, where v is the vector whose entries correspond to the edges from u_1 to the vertices of $\bar{U}$ and similarly for w. Since $v, w, Z_{\bar{G}}(z)$ all have polynomial entries, we conclude that the (u_1, u_2)-entry of $Z_{G'}(z)$ is a rational function. $\square$

Now we give two examples to show that Shannon's algorithm does not literally apply as is to infinite graphs. Let G be a directed VLG with one vertex and $a_k = \lfloor 2^k/(k+1)^2 \rfloor$ edges of length k for all $k \geq 1$. Then $Z_G(z)$ is a 1×1 matrix with sole entry is $f(z) = \sum_k a_k z^k$. In this case we have $f(1/2) \leq (\pi^2/6) - 1 < 1$ and $f(z)$ diverges for any positive $z \geq 1/2$. It is not hard to see that $\lambda_1(G) = 2$ (by Theorem 3.5, since $1 + f + f^2 + \cdots$ converges for $f < 1$, which is the case when $z \leq 1/2$, and clearly diverges when $z > 1/2$ where the series even for f diverges). But the expression $\det(I - Z_G(z))$ fails to have a zero at $z = 1/2$.

Next consider an undirected VLG, G, whose vertex set is the integers, with each vertex having one self loop. Then $\det(I - Z_G(z))$ expanded in a power series has an infinite z coefficient; if we simply multiply the diagonals together (since this is a diagonal matrix), we get the infinite product of $(1 - z)$, which is 0 for any $z > 0$ (yet $\lambda_1(G) = 1$). While G is not connected, we can add edges between i and $i+1$ for all i; this yields a connected graph with similar problems.

Finally we mention that Theorem 3.5 must be modified when G is not strongly connected. Indeed, consider a directed VLG on three nodes, such that the first node has a single self-loop, and the only other edges are edges from the second node to the third, with a_k such edges of length k. Then the most natural way to define $\lambda_1(G)$ is in terms of counting closed walks, so that $\lambda_1(G) = 1$. But if $\sum a_k z^k$ diverges for any $z_0 < 1$, then Theorem 3.5 fails.

5. Limiting Graphs

Let G_i be a sequence of finite VLG's on the same vertex set, V, and the same edge set, E. Let E be partitioned into two sets, E_1, E_2, such that the following holds:
 (1) for each $e \in E_1$, the length of e in G_i is independent of i, and
 (2) for each $e \in E_2$, the length of e in G_i tends to infinity as $i \to \infty$.
The *limit of the* G_i is the graph, G_∞, which is any G_i with its E_2 edges discarded. This simple remark is crucial for an important finiteness lemma (Lemma 9.2).

THEOREM 3.6. *With notation as above,*
$$\lim_{i \to \infty} \lambda_1(G_i) = \lambda_1(G_\infty).$$

PROOF. By counting closed walks we see that $\lambda_1(G_i) \geq \lambda_1(G_\infty)$; this establishes the theorem with "$\geq$" replacing "$=$". To see "$\leq$" replacing "$=$", assume that $\lambda_1\big(A_{G_\infty}(z_0)\big) < 1$ for some $z_0 \in [0,1]$. Clearly $A_{G_i}(z_0) \to A_{G_\infty}(z_0)$ as $i \to \infty$. So the continuity of λ_1 on its entries implies $A_{G_i}(z_0) < 1$ for i sufficiently large, and so $1/z_0 \geq \limsup \lambda_1(G_i)$. Now take a supremum over z_0 with $\lambda_1\big(A_{G_\infty}(z_0)\big) < 1$. □

6. Irreducible Eigenvalues

Let G be an undirected graph with corresponding directed graph $\widehat{G}$. Let G_{Irred} be the graph with vertices $E_{\widehat{G}}$ and an edge from e_1 to e_2 iff $e_1 e_2$ forms an irreducible path in G; i.e., e_1 and e_2 are not opposites (i.e., paired) in G, and e_1 terminates in the vertex where e_2 originates. (Therefore, if e is a half-loop in G, then there is no edge from e to itself in G_{Irred}.) Then walks in G_{Irred} give "irreducible" (or "reduced" or "non-backtracking") walks in G. A closed walk of length k in G_{Irred} gives a closed walk in G (with specified starting vertex) that is *strongly irreducible*, meaning that the closed walk is irreducible and the last step in the closed walk is not the inverse of the first step. We define the *irreducible eigenvalues of* G to be those of G_{Irred}, and we define the largest or Perron-Frobenius eigenvalue of G_{Irred} to be $\lambda_{\text{Irred}} = \lambda_{\text{Irred}}(G)$, the *largest irreducible eigenvalue of* G.

For G an undirected VLG, we may define G_{Irred} as a VLG and hence define $\lambda_{\text{Irred}}(G) = \lambda_1(G_{\text{Irred}})$. In any graph, an irreducible walk that enters a beaded-path must directly traverse this path to its end (any backward step makes the walk reducible). It easily follows that if we subdivide edges in a VLG that are not half-loops (i.e., whole-loops or edges between distinct vertices), then counts of irreducible walks of a given length between V_G vertices remain the same[4]. As a consequence we get the following simple but important proposition.

---

[4]We do not know any simple or very natural way to subdivide half-loops of even length in a VLG while keeping λ_{Irred} invariant. A half-loop of length ℓ should be traversable zero or one time in a row (but not twice or more in a row) in an irreducible walk. For odd lengths, ℓ, this can be achieved by adding a path (of length $(\ell - 1)/2$) with a half-loop at the end. Morally speaking, subdividing an edge in a VLG, G, can be viewed as subdividing the corresponding edge

PROPOSITION 3.7. *We have λ_{Irred} of a VLG is invariant under the subdivision of any set of edges devoid of half-loops; in particular, λ_{Irred} is invariant under passing to a subdivided form of any VLG with all half-loops of length one. Similarly, λ_{Irred} of a graph is invariant under any suppression.*

We now state a theorem for use later in this paper; the theorem requires a definition.

DEFINITION 3.8. *A connected graph, G, is loopy if $|E_G| \geq |V_G|$, or equivalently if G contains an irreducible closed walk, or equivalently if G is not a tree. G is 1-loopy if G is connected and the removal of any edge from G leaves a graph each of whose connected components are loopy.*

THEOREM 3.9. *For a connected graph, G, the following are equivalent:*
(1) *G is 1-loopy,*
(2) *G_{Irred} is strongly connected, and*
(3) *G is not a cycle and all vertices in G have degree at least 2.*

In particular, if G is connected and d-regular for $d \geq 3$, then G_{Irred} is strongly connected.

Condition (3) in the above theorem was pointed out to us by a referee.

PROOF. (2)⇒(1): Consider a directed edge, $e \sim (u, v)$, of G, and let e' be e's opposite (we permit $e = e'$, i.e., the case of a half-loop). If the removal of e leaves the connected component of v being a tree, then there is no irreducible walk from e to e'. On the other hand, if this connected component is not a tree, then a cycle in this component about v of minimum length is irreducible, which then extends to an irreducible walk from e to e'. To summarize, if G is not 1-loopy, then G_{Irred} is not strongly connected; otherwise, each edge has an irreducible path to its opposite, i.e., each edge is connected to its opposite in G_{Irred}.

(1)⇒(2): Assume G is 1-loopy. If e_1, e_2 are two distinct, unpaired directed edges originating in the same vertex, v, then an irreducible path from e_1 to its opposite followed by e_2 shows that e_1 and e_2 are connected by an irreducible path. Thus any two edges that share a vertex are strongly connected in G_{Irred}.

Finally if e_1, e_2 are two undirected edges without a common vertex, then a shortest path connecting them gives an irreducible path from some orientation of e_1 to some of e_2. By the above we can follow them with irreducible paths to the edge of opposite orientation. Thus any two edges that do not share a vertex are strongly connected in G_{Irred}. So G_{Irred} is strongly connected.

(1)⇒(3) Since an isolated vertex is loopy, a 1-loopy graph has all vertices of degree at least 2. Also a 1-loopy graph cannot be a single cycle, since removing any one edge would give a tree.

(3)⇒(1) Any tree either (i) is an isolated vertex, or (ii) is a path with two vertices of degree one, or (iii) has at least three vertices of degree 1 (this follows by considering two vertices, u, v, of maximum distance in a tree, and then taking a vertex of maximum distance to the unique path joining u to v). Consider the tree,

---

pair in underlying directed graph, $\widehat{G}$, and then gluing them together with an "opposite" pairing (that glues the newly introduced vertices together); for half-loops, we should glue a single directed self-loop of length ℓ to itself, via reflection about the middle); for even length half-loops, that fact that reflection leaves the middle vertex fixed creates problems if we wish to remain in the category of undirected graphs...

T, obtained as the non-loopy connected component by removing an edge, e, from a graph G that is non-loopy. If T is the only connected component, e has both its endpoints in T, and if not then T has only one endpoint in T; in either situation, considering the cases (i)–(iii), it is easy to check that G is a cycle or has at least one vertex (a vertex in T) of degree 1. □

7. λ_1 and Closed Walks for Infinite Graphs

In this subsection we recall some facts about λ_1 and closed walks of graphs, either indicating the proofs or giving references. This section is geared to infinite graphs, since most of the facts are very easy when the graph is finite.

Let G be a graph of bounded degree, i.e. there is an r such that the degree of each vertex is at most r. Then A_G, G's adjacency matrix, is a bounded linear operator (bounded by r) on $L^2(G)$, the square summable functions on G's vertices. The next theorem shows that $\|A_G\|$, the norm of the operator $\|A_G\|$, equals $\lambda_1(G)$ as was defined by counting closed walks about a vertex of at most some given length.

THEOREM 3.10. *Assume G is connected. For any vertex, v, of G and positive integer, k, recall that $c(v,k)$ is the number of closed walks of length k in G about v. Then $c(v,k) \le \|A_G\|^k$, and*

$$\lim_{r \to \infty} [c(v, 2r)]^{1/(2r)}$$

exists and equals $\|A_G\|$.

Since the limit above equals $\lambda_1(G)$, we see that $\lambda_1(G) = \|A_G\|$.

PROOF. See [**Buc86**]. □

We also need the following simple fact.

THEOREM 3.11. *For every $\epsilon > 0$ there is an $f \ne 0$ such that $\|Af\| \ge (\lambda_1 - \epsilon)\|f\|$, where f has finite support.*

PROOF. By definition of norm, there is a $y \ne 0$ with $\|Ay\| \ge (\lambda_1 - (\epsilon/2))\|g\|$. For any $\nu > 0$ we may write $g = g_1 + g_2$ where g_1 is of finite support and $\|g_2\| \le \nu$. It is easy to see (using the fact that A is bounded) that if ν is sufficiently small we can take $f = g_1$ to satisfy the above theorem. □

8. A Curious Theorem

DEFINITION 3.12. *Let ψ be a finite connected graph with each vertex of degree at most d for an integer $d > 2$. By $\text{Tree}_d(\psi)$ we mean the unique (up to isomorphism) undirected graph, G, that has an inclusion $\iota: \psi \to G$ such that G is d-regular and such that G becomes a forest when we remove (the image under ι of) ψ's edges.*

We have seen an example of this construction in Chapter 2, where ψ is one vertex with m self-loops, in the proof of Theorem 2.11. See Figure 1 for another example.

In the category of d-regular graphs with a ψ inclusion, $\text{Tree}_d(\psi)$ is none other than the universal cover.

The methods used to prove Theorems 1.1, 1.2, and 1.3 suggest the following curious theorem.

8. A CURIOUS THEOREM

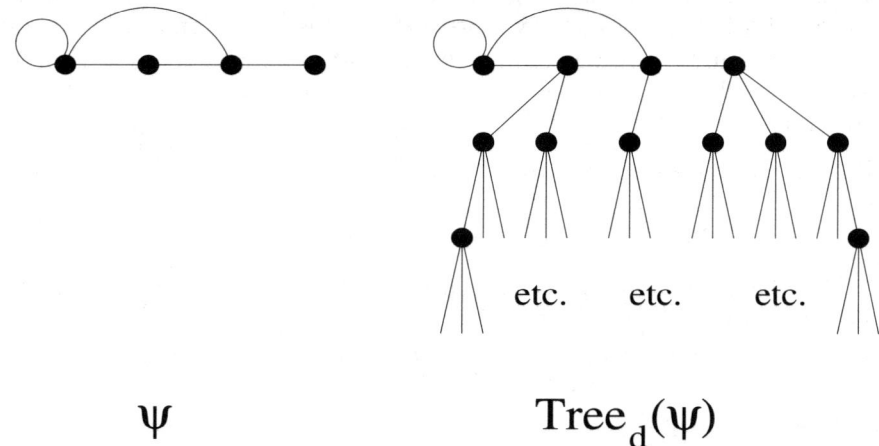

ψ \hspace{4cm} Tree$_d(\psi)$

FIGURE 1. A graph, ψ, and Tree$_d(\psi)$ with $d = 4$.

THEOREM 3.13. *Let $d \geq 3$, and let ψ be a finite connected graph with each vertex of degree $\leq d$. Then*

$$\lambda_1\big(\text{Tree}_d(\psi)\big) = 2\sqrt{d-1} \iff \lambda_{\text{Irred}}(\psi) \leq \sqrt{d-1},$$
$$\lambda_1\big(\text{Tree}_d(\psi)\big) > 2\sqrt{d-1} \iff \lambda_{\text{Irred}}(\psi) > \sqrt{d-1}.$$

The same is true for any real $d > 2$, provided that $\lambda_1\big(\text{Tree}_d(\psi)\big)$ is interpreted with an appropriate analytic continuation in d (described below).

Before proving the theorem, we describe the analytic continuation to which we refer.

Let $\widetilde{T}$ be an undirected rooted tree, where every vertex has $d-1$ children, with $d > 1$ an integer (for now). ($\widetilde{T}$ has degree $d-1$ at the root and degree d elsewhere, so $\widetilde{T}$ is not regular.) Let a_n for $n = 2, 4, \ldots$ be the number of walks in $\widetilde{T}$ from the root to itself that never pass through the root except at the beginning and end. It is easy to see that

$$S = S_d(z) = \sum_{n=2}^{\infty} a_n z^n,$$

satisfies the recurrence $S = z^2(d-1)(1 + S + S^2 + \cdots)$ (since the walks counted by S take one step, in $d-1$ possible ways, to a child of the root, followed by some number (possibly zero) of S walks, followed by the step back to the root; so $S(1-S) = z^2(d-1)$, so that near $z = 0$

(13) $$S = S_d(z) = \frac{1 - \sqrt{1 - 4(d-1)z^2}}{2}.$$

One can alternatively say that if H is the realization of $\widetilde{T}$ with vertex set $\{v_0\}$, where v_0 is the root, then the 1×1 matrix $Z_H(z)$ has entry $S(z)$. Similarly, if $\widetilde{T}$ were modified to have $d-r$ children at the root (but every other vertex with $d-1$ children, as before), then $Z_H(z)$ would have entry $(d-r)S(z)/(d-1)$.

By $\lambda_1\big(\text{Tree}_d(\psi)\big)$ for $d > 1$ real (assuming each degree in ψ is $\leq d$) we mean λ_1 of the VLG formed from ψ with an additional self-loop about each vertex of degree

r that has "formal weight" $(d-r)S_d(z)/(d-1)$ (this can be viewed as adding an infinite set of self-loops of given weights and lengths corresponding to this power series[5], or can simply be viewed as a term to add to the diagonal of $Z_G(z)$). When d in an integer, the realization of $\text{Tree}_d(\psi)$ with vertex set V_ψ is exactly this VLG, and by Proposition 3.7 we know λ_{Irred} remains the same.

PROOF. By [**God93**], exercise 13 page 72, we know that $\lambda_{\text{Irred}}(A)$ is given by $1/y$ of the smallest root, y, of
$$\det(I - yA + y^2(D-I)) = 0,$$
where D is the diagonal matrix whose entry at vertex v is the degree of v, and I is the identity matrix. Now
$$Z_G(z) = zA + \frac{dI - D}{d-1} S(z).$$
It is then easy to verify that
$$I - Z_G(z) = (1 - S(z))(I - yA + y^2(D-I))$$
for $y = y(z) = z/(1 - S(z))$ (note that $y(z)$ is an increasing function in z for as long as $z(1 + S + S^2 + \cdots)$ converges). But $S(z_1) \leq 1/2$ for all positive real $z_1 \leq z_0$, where $z_0 = 1/(2\sqrt{d-1})$, and also $y(z_0) = 1/\sqrt{d-1}$. So $I - y_1 A + y_1^2(D-I)$ becomes non-invertible for a $y_1 < y(z_0)$ precisely when $Z_G(z_1)$ has eigenvalue 1 for a $z_1 \in [0, z_0]$. Now apply the equality of quantities (1) and (3) in Theorem 3.5. □

Note that the proof shows that if $1/y_1 = \lambda_{\text{Irred}}(A)$ and $1/z_1$ is $\text{Tree}_d(\psi)$, then $y_1 = y(z_1)$ provided $1/z_1 > 2\sqrt{d-1}$.

---

[5]In other words, we view a power series $\sum a_n z^n$ as the sum of terms representing a_n edges of length n (even when a_n is not an integer).

CHAPTER 4

Tangles

In Chapter 2 we saw that a vertex with m self-loops in a $\mathcal{G}_{n,d}$ graph, with m "large," gives rise to a "large" second eigenvalue (i.e., larger than $2\sqrt{d-1}$ for sufficiently large m, as $n \to \infty$). Here we generalize this observation to what we call a "tangle." Our proof of the Alon conjecture via a trace method must somehow overcome all "hypercritical" tangles.

DEFINITION 4.1. *Given two Π-labelled graphs, G and H, we say G contains H (or H occurs in G) if there is an inclusion*[1] *$\iota\colon H \to G$ that preserves the labelling; the number of times a graph, H, occurs in G is the number of distinct*[2] *such ι. A tangle (or $\mathcal{G}_{n,d}$-tangle) is a Π-labelled connected graph, ψ, that is contained in some element of $\mathcal{G}_{n,d}$.*

For example, in Chapter 2 we studied the tangle with one vertex and m self-loops labelled $\pi_1, \ldots, \pi_m$. We define $\mathcal{H}_{n,d}$- and $\mathcal{I}_{n,d}$- and $\mathcal{J}_{n,d}$-tangles similarly, with the following modifications to the meaning of the π_i's. For $\mathcal{H}_{n,d}$, each of the $d/2$ independent π_i's is uniform over all permutations whose cyclic decomposition consists of a single cycle. For $\mathcal{I}_{n,d}$, each of d independent π_i's are uniform over all perfect matchings, i.e., over all permutations that are involutions without fixed points. For $\mathcal{J}_{n,d}$, each of d independent π_i's are uniform over all involutions with exactly one fixed point.

THEOREM 4.2. *Fix a positive integer, $d \geq 3$, and a graph ψ. Any graph, G, on n vertices, that contains ψ has second eigenvalue at least $\rho - o(1)$, where $o(1)$ is a function of n tending to 0 as $n \to \infty$, and where ρ is the norm of the adjacency matrix of $\mathrm{Tree}_d(\psi)$.*

This generalizes Theorem 2.11.

PROOF. Fix $\epsilon > 0$; we will show that for n sufficiently large any such G has $\lambda_2(G) \geq \rho - \epsilon$. First, there is a finitely supported $f \neq 0$ on $\mathrm{Tree}_d(\psi)$ with $\|Af\| \geq \|f\|(\rho - \epsilon)$, where A is the adjacency matrix of $\mathrm{Tree}_d(\psi)$. If V_ψ is the set of vertices on which f is non-zero, then replacing f with the non-negative first Dirichlet eigenfunction on V_ψ (see [**Fri93**]) we may assume f is non-negative, nonzero, and that $Af \geq f(\rho - \epsilon)$ (with equality everywhere except at the boundary of V_ψ). There is a covering map (see [**Fri93**] and [**Fri03**]) $\pi\colon \mathrm{Tree}_d(\psi) \to G$; set $\pi_* f$ to be

---

[1] By an inclusion we mean a graph homomorphism that is a injection on the vertices and on the edges.

[2] For example, if $H = G$ consists of one edge joining two distinct vertices, u, v, then the identity is considered distinct from the morphism interchanging the vertices. By the same principle, if H has exactly k automorphisms, then the number of times H occurs in a graph is always a multiple of k.

the function on G defined by
$$(\pi_* f)(v) = \sum_{\pi(w)=v} f(w).$$
If A_G is the adjacency matrix of G, then clearly
$$A_G(\pi_* f) \geq (\rho - \epsilon)(\pi_* f).$$
So, on the one hand, $\mathcal{R}(\pi_* f) \geq \rho - \epsilon$, where $\mathcal{R}$ is the Rayleigh quotient for A_G. On the other hand, the support, N, of $\pi_* f$ is of size no greater than that of f, and this size is bounded (independent of G and n). So the same reasoning as in the proof of Theorem 2.11 shows that
$$\mathcal{R}(g) \geq d - o(1), \qquad \text{where} \quad g = \chi_{V \setminus \tilde{N}},$$
where $\tilde{N}$ is the set of vertices of distance 0 or 1 to N. Since $\pi_* f$ is orthogonal to both g and $A_G g$, we are done (by Lemma 2.9). $\square$

DEFINITION 4.3. *A tangle, ψ, is* critical *(respectively,* supercritical, hypercritical*) if $\lambda_{\text{Irred}}(\psi)$ equals (respectively, is at least, exceeds) $\sqrt{d-1}$.*

According Theorems 3.13 and 4.2, a fixed hypercritical tangle can only occur in a graph with sufficiently many vertices if the graph has $\lambda_2 > 2\sqrt{d-1}$.

Now that we know how tangles affect eigenvalues, we want to know how often the tangles occur. This discussion, and the particular application to $\mathcal{H}_{n,d}$, $\mathcal{I}_{n,d}$, and $\mathcal{J}_{n,d}$, will take the rest of this section.

DEFINITION 4.4. *A* leaf *on a graph is a vertex of total degree 1 (whether the graph is directed or not). We say that a graph is* pruned *if it has no leaves. A* simple pruning *is the act of removing one leaf and its incident edge from a graph;* pruning *is the repeated performance of some sequence of simple prunings;* complete pruning *is the act of pruning until no more pruning can be done.*

For example, completely pruning a tree results in a single vertex with no edges; completely pruning a cycle leaves the cycle unchanged.

PROPOSITION 4.5. *Given a graph, G, there is a unique pruned graph H obtainable from completely pruning G. Furthermore, H is completely pruned iff each edge of H lies on an irreducible cycle.*

PROOF. Let $e_1, \ldots, e_t$ denote the edges pruned in one pruning of G, in the order in which they are pruned. Let G' be a different complete pruning of G, which we assume does not contain all the e_i. Let j be the smallest integer such that e_j lies in G'. On the one hand, the removal of $e_1, \ldots, e_{j-1}$ from G, or any subgraph of G, allows e_j to be pruned from G, or any subgraph of G, including G'. On the other hand, the prunability of e_j from G' contradicts the completeness of the pruning that formed G'. It follows that any complete pruning contains all the edges of any other, and so any two are the same.

We now address the last statement of the theorem. If H has a leaf, then the edge incident upon this leaf does not lie in an irreducible cycle. Conversely, if H has no leaves, and if e is an edge with endpoints u, v, consider the graph, H', obtained by removing e from H. If u and v are connected in H', then a minimal length path that joins them, along with e, gives an irreducible cycle containing e. Otherwise, since u's connected component is not a tree (or else H would have leaves), this component has an irreducible cycle, and a shortest walk from u to this cycle, once around the cycle, and back to u gives an irreducible cycle beginning and ending at u. Similarly there is such a cycle about v. The cycle about u, followed by e, followed by the w cycle, and back through e (in the other direction), gives an irreducible cycle containing e. □

A morphism of tangles is a morphism of Π-labelled graphs, i.e., a graph morphism that preserves the edge labelling.

DEFINITION 4.6. *The* order *of a tangle*, ψ, *is* $\operatorname{ord}(\psi) = |E_\psi| - |V_\psi|$ *(so while a whole-loop is counted as one edge, for the model $\mathcal{J}_{n,d}$, to be considered soon, a half-loop is also counted as one edge[3]). More generally, for a graph, G, or any structure with an underlying graph, G (such as a "form" or "type" to be defined in Chapter 5), its order is* $\operatorname{ord}(G) = |E_G| - |V_G|$.

THEOREM 4.7. *Let ψ be a tangle of non-negative order. Then the expected number of occurrences of ψ in an element, G, of $\mathcal{G}_{n,d}$ is $n^{-r} + O(n^{-r-1})$, where r is the order of ψ. The probability that at least one occurrence occurs is at least $n^{-r}/c - n^{-2r}/(2c^2) + O(n^{-r-1})$, with r as before and where c is the number of automorphisms of the complete pruning of ψ.*

Notice that the number of automorphisms of the complete pruning of a tangle, ψ, is at least as many as that of ψ, and it may be strictly greater[4]. The proof following will imply that the probability of ψ's occurrence is also at most $n^{-r}/c + O(n^{-r-1})$, which matches the lower bound to first order, provided that $r \geq 1$.

PROOF. Let $V_\psi = \{u_1, \ldots, u_s\}$, and for a tuple $\vec{m} = (m_1, \ldots, m_s)$ of distinct integers between 1 and n, let $\iota_{\vec{m}}$ denote the event that the map, ι, mapping u_i to m_i, is an occurrence of ψ in G. If a_i is the number of ψ's edges labelled π_i, then each event $\iota_{\vec{m}}$ involves setting a_i values of π_i, all of which occur with probability

$$(14) \qquad \frac{(n-a_1)!}{n!} \cdots \frac{(n-a_{d/2})!}{n!}.$$

Since the sum of the a_i is $|E_\psi|$, this probability is $n^{-|E_\psi|}$. Since there are $n!/(n-|V_\psi|)! = n^{|V_\psi|} + O(n^{|V_\psi|-1})$ different $\iota_{\vec{m}}$'s, the expected number of occurrences is $n^{-r} + O(n^{-r-1})$, where $r = \operatorname{ord}(\psi)$.

Next notice that if ψ' is a pruning of a tangle, ψ, then the probability that ψ occurs is $1 + O(n^{-1})$ times the probability that ψ' occurs (adding each pruned edge adds a condition that occurs with probability between 1 and $(n-c_1)/(n-c_2)$ with c_1, c_2 constants). Hence we may assume ψ is pruned.

An automorphism of ψ can be viewed as a permutation on V_ψ, which is the same as a permutation, σ, on $\{1, \ldots, s\}$ (identifying a $u_i \in V_\psi$ with i). Such a permutation, σ, acts by permuting the components of the $\vec{m}$'s. Say that $\iota_{\vec{m}}$ is equivalent to $\iota_{\vec{k}}$ if $\vec{m}$ and $\vec{k}$ differ by a permutation, σ, associated to an automorphism of ψ; i.e., if $\iota_{\vec{m}}$ and $\iota_{\vec{k}}$ correspond to the same subgraph of G. Let R be a set of representatives in the equivalence classes of all $\vec{m}$'s.

---

[3]Rougly speaking, the reason for this is that a whole-loop and half-loop are both $1/n + O(1/n^2)$ probability events.

[4]Indeed, a structural induction argument (i.e., by pruning one leaf) shows that any automorphism of the complete pruning of ψ has at most one extension to ψ. On the other hand, if ψ is a cycle of length q with all edges in one "direction" labelled π_1, then ψ has q automorphisms; yet if we add one edge labelled π_2 to ψ at any vertex, the new graph has only the trivial automorphism.

By inclusion/exclusion, the probability that ψ occurs at least once is at least

$$\sum_{\vec{m} \in R} \text{Prob}\{\iota_{\vec{m}}\} - \frac{1}{2} \sum_{\substack{\vec{k} \neq \vec{m} \\ \vec{k}, \vec{m} \in R}} \text{Prob}\{\iota_{\vec{m}} \cap \iota_{\vec{k}}\}. \tag{15}$$

The first summand is $(1/c)n^{-r} + O(n^{-r-1})$, by the argument given for the expected number. For the second summand, we may write $\iota_{\vec{m}} \cap \iota_{\vec{k}}$ as $\iota_{\vec{q}}(\psi')$, where $\vec{q}$ is a vector comprised of the distinct components of $\vec{m}$ and $\vec{k}$, and where ψ' is the tangle obtained by gluing two copies of ψ along certain vertices (corresponding to where the components of $\vec{m}$ and $\vec{k}$ coincide). If $\vec{m}$ is disjoint from (i.e., nowhere coincides with) $\vec{k}$, then ψ' is two disjoint copies of ψ; for fixed $\vec{k} \in R$ we have

$$\sum_{\substack{\vec{m} \in R \\ \vec{m} \text{ disjoint from } \vec{k}}} \text{Prob}\{\iota_{\vec{m}} \mid \iota_{\vec{k}}\} = n^{-r}/c + O(n^{-r-1}),$$

the summation being over conditional probabilities, since the conditioning of $\iota_{\vec{k}}$ and summing over $\vec{m}$ disjoint only affects equation (14) by changing $n!/(n-a_i)!$ terms into $(n-c_i)!/(n-c_i-a_i)!$ terms for constants c_i, which is a second order change. Hence

$$\sum_{\substack{\vec{k}, \vec{m} \in R \\ \vec{k}, \vec{m} \text{ disjoint}}} \text{Prob}\{\iota_{\vec{m}} \cap \iota_{\vec{k}}\} = \sum_{\vec{k} \in R} \text{Prob}\{\iota_{\vec{k}}\} \sum_{\substack{\vec{m} \in R \\ \vec{m} \text{ disjoint from } \vec{k}}} \text{Prob}\{\iota_{\vec{m}} \mid \iota_{\vec{k}}\}$$

$$= \sum_{\vec{k} \in R} \text{Prob}\{\iota_{\vec{k}}\} \left(n^{-r}/c + O(n^{-r-1})\right) = n^{-2r}/c^2 + O(n^{-2r-1}).$$

To understand the situation where $\vec{m}$ and $\vec{k}$ overlap somewhere, we pause for some lemmas.

LEMMA 4.8. *Let $\iota: \psi \to G$ be an inclusion of graphs, with G connected. Then the order of G is at least that of ψ.*

PROOF. Let G_ψ be G with the vertices of $\iota(\psi)$ identified, and all $\iota(\psi)$ edges discarded. Then G_ψ is connected, and so has order at least -1; on the other hand, clearly the order of G_ψ is the order of G minus that of ψ minus 1 (for the vertex that is the identification of all $\iota(\psi)$ vertices). Hence

$$\text{ord}(G) = \text{ord}(G_\psi) + \text{ord}(\psi) + 1 \geq -1 + \text{ord}(\psi) + 1 = \text{ord}(\psi).$$

□

LEMMA 4.9. *Let G be a pruned graph and let $e \in E_G$. If $G \setminus \{e\}$ (i.e., G with e removed) has two connected components, then each connected component has order at least 0.*

PROOF. Consider a connected component, G', of $G \setminus \{e\}$. If G' did not contain a cylce, then G' would be a tree, and then G would not be completely pruned. So G' contains a cycle, and we may apply Lemma 4.8 to deduce that G' has order at least 0. □

LEMMA 4.10. *Let $\iota: \psi \to G$ be an inclusion of pruned graphs. Then the order of G is at least that of ψ, and G's order is strictly greater than ψ's if $\iota(\psi)$ is properly contained in G.*

PROOF. A connected component of a graph that is pruned (and non-empty) has non-negative order. So we may assume $\iota(\psi)$ meets every connected component of G. It suffices to prove the case where $\iota(\psi)$ meets one connected component of G, i.e., the case where G is connected. Choosing an edge, e, that is missed by $\iota(\psi)$, we have that ι includes ψ into $G \setminus \{e\}$ (i.e., G with e removed); we apply Lemma 4.8 to those components of $G \setminus \{e\}$ containing part of $\iota(\psi)$, and to the possibly one other component of $G \setminus \{e\}$ we apply Lemma 4.9. It follows that the order of ψ is at most that of $G \setminus \{e\}$; but the order of $G \setminus \{e\}$ is one less than that of G. □

LEMMA 4.11. *Let $\iota_1, \iota_2 \colon \psi \to H$ be two inclusions of a tangle, ψ, in a connected (labelled) graph, H, such that $\iota_1(\psi) \cup \iota_2(\psi) = H$. Assume that $\iota_1(\psi) \neq H$. Then the order of H is greater than that of ψ.*

PROOF. Ignoring ι_2, the preceding lemma applies to ι_1 to immediately yield this lemma. □

Lemma 4.11 shows that

$$\sum_{\substack{\vec{k},\vec{m} \text{ not disjoint} \\ \vec{k},\vec{m} \in R, \ \vec{k} \neq \vec{m}}} \mathrm{Prob}\{\iota_{\vec{m}} \cap \iota_{\vec{k}}\} = O(n^{-r-1}),$$

since the summation can be broken down into a finite number of sums over tangles of order at least $r+1$. Thus

$$\sum_{\vec{m} \in R} \mathrm{Prob}\{\iota_{\vec{m}}\} - \frac{1}{2} \sum_{\substack{\vec{k} \neq \vec{m} \\ \vec{k},\vec{m} \in R}} \mathrm{Prob}\{\iota_{\vec{m}} \cap \iota_{\vec{k}}\}.$$

$$= n^{-r}/c + O(n^{-r-1}) - n^{-2r}/(2c^2) + O(n^{-2r-1}) + O(n^{-r-1}),$$

which completes the proof of Theorem 4.7. □

It is easy to see that the above proof of Theorem 4.7 uses very little about the model of random graph, and therefore generalizes as follows.

THEOREM 4.12. *Let $\mathcal{K}_n$ be a model of d-regular random graphs on n vertices labelled $\{1, \ldots, n\}$, defined for some values of n. Further assume that (1) $\mathcal{K}_n$ is invariant under renumbering $\{1, \ldots, n\}$, and (2) any tangle, ψ, has expected number of occurrences $n^{-r} + O(n^{-r-1})$ where r is the order of ψ. Then Theorem 4.7 holds for $\mathcal{K}_n$.*

THEOREM 4.13. *For G drawn from $\mathcal{H}_{n,d}$ or $\mathcal{I}_{n,d}$, we have that $\lambda_2(G) > 2\sqrt{d-1}$ with probability at least $n^{-s}/2 + O(n^{-s-1})$ where $s = \lfloor\sqrt{d-1}\rfloor$, except for when $d = 4$ in $\mathcal{H}_{n,d}$, where we may take $s = 2$. The same holds for $\mathcal{J}_{n,d}$ with probability $n^{-s} + O(n^{-s-1})$, where $s = \lfloor(\sqrt{d-1}+1)/2\rfloor$ (and no exceptional values of d).*

PROOF. For $\mathcal{J}_{n,d}$, the tangle consisting of 1 vertex with a number of self-loops (in this case half-loops), proves the theorem for $\mathcal{J}_{n,d}$ just as it did for $\mathcal{G}_{n,d}$ in Theorem 2.11.

Consider the tangle, ψ, with two vertices and m edges joining the two vertices labelled $\pi_1, \pi_2, \ldots$; ψ is an $\mathcal{H}_{n,d}$-tangle provided that $m \leq d/2$, and an $\mathcal{I}_{n,d}$-tangle if $m \leq d$. $\lambda_{\mathrm{Irred}}(\psi)$ is clearly $m-1$, $s = \mathrm{ord}(\psi) = m-2$, and the automorphism group of ψ is of order 2. So if $m - 1 > \sqrt{d-1}$ and if G contains this tangle, then

we have $\lambda_2(G) > 2\sqrt{d-1}$ for n sufficiently large. If we take $s = \lfloor\sqrt{d-1}\rfloor$, then $m - 1 > \sqrt{d-1}$; we require $m \leq d$ for $\mathcal{I}_{n,d}$, amounting to
$$\lfloor\sqrt{d-1}\rfloor + 2 \leq d,$$
which is satisfied for all $d \geq 3$. For $\mathcal{H}_{n,d}$ we require
$$\lfloor\sqrt{d-1}\rfloor + 2 \leq d/2,$$
which is satisfied for all $d \geq 7$. Since d is even and ≥ 4 in $\mathcal{H}_{n,d}$, we finish by examining the cases $d = 4$ and $d = 6$.

For $d = 4$ consider the tangle, ψ, with vertices v_1, v_2, v_3, v_4, edges labelled π_1, π_2 from v_1 to v_2, from v_2 to v_3, and from v_3 to v_4. Then $\text{ord}(\psi) = 2$ and $\lambda_{\text{Irred}}(\psi) > \lambda_{\text{Irred}}(\psi')$ where ψ' is the subgraph of ψ induced on v_1, v_2, v_3. But in the proof of Theorem 6.10 we compute $\lambda_{\text{Irred}}(\psi') = \sqrt{3}$. Hence ψ is hypercritical of order 2, so we may take $s = 2$ in the theorem in the case of $d = 4$ and $\mathcal{H}_{n,d}$.

For $d = 6$, the proof of Theorem 6.10 gives a tangle of order 2 with $\lambda_{\text{Irred}} > \sqrt{5}$ (see equation (30) and the discussion around it). So we may take $s = 2$ in our theorem when $d = 6$ in $\mathcal{H}_{n,d}$.

$\square$

CHAPTER 5

Walk Sums and New Types

In this section we give some general techniques that are used in estimating the expected values of all the various traces that are used in this paper. The main idea, originated in [**BS87**] and strengthened in [**Fri91**], is to group contributions to the trace in the following way. Consider the word, w, and vector, t

$$w = (\sigma_1, \ldots, \sigma_{10}) = (\pi_2^{-1}, \pi_1, \pi_3, \pi_1^{-1}, \pi_3, \pi_3, \pi_1^{-1}, \pi_3, \pi_1^{-1}, \pi_2),$$

$$\vec{t} = (t_0, \ldots, t_{10}) = (5, 2, 4, 3, 7, 4, 3, 7, 4, 2, 5)$$

(see Figure 1). This represents a possible or potential irreducible closed walk of length 10, from 5 to 2 along the edge labelled π_2^{-1}, from 2 to 4 along π_1, etc. Graphically we depict this potential walk by the subgraph it traces out, called its "form" (see Figure 1). The pair $(w; \vec{t})$ represents a possible contribution to IrredTr $(A, 10)$. For any word, w, of length k and vector, $\vec{t}$, of length $k+1$, let $P(w; \vec{t})$ be the probability that $\sigma_i(t_{i-1}) = t_i$ for all i, i.e., that the cycle does occur. Then the expected value of IrredTr $(A, 10)$ is just a sum of appropriate $P(w; \vec{t})$'s, what we will call a "walk sum."

In the form of Figure 1, we see that the numbers of the vertices $5, 2, 4, 3, 7$, are irrelevant, since our random graph model is "symmetric," or invariant under renumbering the vertices. So we may replace the vertex 5 by an abstract symbol v_1, 2 by v_2, etc. (for reasons to made clear later, we do want to remember the order

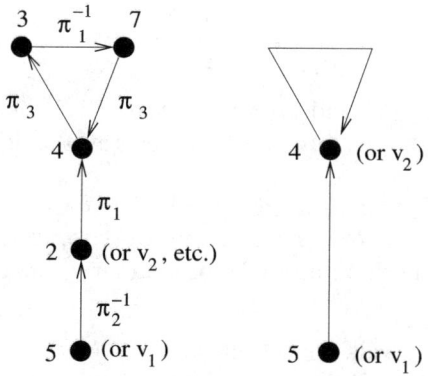

A Form   Its Type

FIGURE 1. A form and its associated type.

in which the vertices were traversed); this replacement is not necessary, but makes clear that there is no particular preference for any numbering of the vertices.

In our example of Figure 1, $(w; \vec{t})$, a loop about the vertex 4 is traversed twice before we return to the starting vertex 2. Broder and Shamir realized that the other vertices, those of degree 2 that are not the starting vertex (here the vertices 2, 3, 7) are less interesting features of the "form." By suppressing these "less interesting" vertices we get the "type" of the form (see Figure 1); in [**Fri91**], walk sums were grouped by their form, and the sums for each form were grouped by the type of the form. A type is a graph with certain features, but when a form gives rise to a type then the edges of the type inherit Π^+ labels from the form, and inherit "lengths," which are the lengths of the Π^+ labels (recall that Π^+ is the set of nonempty words over Π). For example, the edge (5, 4) of the type in Figure 1 inherits a length of 2 and a label of $\pi_2^{-1}\pi_1$ from the form next to it; edge (4, 4) inherits a length of 3 and a label of $\pi_3\pi_1^{-1}\pi_3$.

This paper introduces a "new type," used in analyzing walk sums corresponding to selective traces. A new type is a type with some additional information, primarily fixing the lengths of certain type edges and requiring all other lengths to be sufficiently large.

1. Walk sums

By a *weakly potential (k, n)-walk*, $(w; \vec{t})$, we mean a pair consisting of a word $w = \sigma_1 \ldots \sigma_k$ of length k over Π, and a vector, $\vec{t} = (t_0, \ldots, t_k)$, with each $t_i \in \{1, \ldots, n\}$; we sometimes refer to k as the *length* and n as the *size* of the weakly potential walk. Given such a $(w; \vec{t})$, let $\mathcal{E}(w; \vec{t})$ denote the event that the π_i are chosen so that $\sigma_i(t_{i-1}) = t_i$ for all $i = 1, \ldots, k$. Let $P(w; \vec{t})$ denote the probability that $\mathcal{E}(w; \vec{t})$ occurs.

A *potential (k, n)-walk* is a weakly potential (k, n)-walk that is *feasible*, meaning that $P(w; \vec{t}) \neq 0$, or equivalently that the following two feasibility conditions hold: (1) for any i, j such that $\sigma_i = \sigma_j$, we have

$$t_{i-1} = t_{j-1} \Leftrightarrow t_i = t_j,$$

and for any i, j such that $\sigma_i = \sigma_j^{-1}$, we have

$$t_{i-1} = t_j \Leftrightarrow t_i = t_{j-1}.$$

For example, if $\vec{t} = (1, 1, 2)$ and $w = \pi_1\pi_1$, then $(w, \vec{t})$ is a weakly potential walk but not a potential walk. A *potential (k, n)-closed walk* is a potential walk, $(w, \vec{t})$ as above with $t_k = t_0$.

All our variants of traces, irreducible and not irreducible, selective and not, can be viewed as sums of $P(w; \vec{t})$ over appropriate w's and $\vec{t}$'s. In this subsection we formalize this notion and make preliminary remarks about such sums and asymptotic expansions.

DEFINITION 5.1. *A walk collection, $\mathcal{W} = \mathcal{W}(k, n)$, is a collection, for any two positive integers k and n, of (k, n)-potential walks $(w; \vec{t})$ as above, i.e. w is a word over Π of length k, and $\vec{t} = (t_0, \ldots, t_k)$ is a $k+1$ dimensional vector over $\{1, \ldots, n\}$, and $(w; \vec{t})$ is feasible. The* walk sum *associated to $\mathcal{W}$ is*

$$\text{WalkSum}(\mathcal{W}, k, n) = \sum_{(w;\vec{t}) \in \mathcal{W}(k,n)} P(w; \vec{t}).$$

The main goal of this paper is to organize the various $(w, \vec{t})$ pairs into groups over which we can easily sum $P(w; \vec{t})$. One simple organizational remark is that symmetries in the $(w; \vec{t})$ pairs often simplify matters. Specifically, given an permutation, τ, of the integers, and a vector $\vec{t}$ as above, let

$$\tau(\vec{t}) = (\tau(t_0), \ldots, \tau(t_k)).$$

Say that $\vec{s}$ and $\vec{t}$ *differ by a symmetry* if $\vec{s} = \tau(\vec{t})$ for some τ; in this case clearly $P(w, \vec{s}) = P(w, \vec{t})$ for any word w of length k. We use $\vec{s} \sim \vec{t}$ to denote that $\vec{s}$ and $\vec{t}$ differ by a symmetry.

DEFINITION 5.2. *A walk collection, $\mathcal{W}$, is* SSIIC *if it is*
(1) *symmetric, i.e. $(w, \vec{t}) \in \mathcal{W}(k, n)$ implies that $(w, \vec{s}) \in \mathcal{W}(k, n)$ for all $\vec{s} \sim \vec{t}$ such that $s_i \le n$ for all i,*
(2) *size invariant, i.e. if $(w, \vec{t})$ is a potential (k, n)-walk, then for any $n' > n$, $(w, \vec{t}) \in \mathcal{W}(k, n)$ iff $(w, \vec{t}) \in \mathcal{W}(k, n')$,*
(3) *irreducible, meaning that $(w, \vec{t}) \in \mathcal{W}$ implies that w is irreducible, and*
(4) *closed, meaning that $(w, \vec{t}) \in \mathcal{W}$ implies that $t_0 = t_k$.*

The walk sums of interest here, namely traces that are irreducible or strongly irreducible and possibly selective, will all be SSIIC. We now make a series of remarks about walk sums that apply to all SSIIC walk sums; some of the remarks apply more generally.

DEFINITION 5.3. *Let $\vec{t}$ be as above, i.e., a positive integer valued vector. Define the* equivalence class of $\vec{t}$ *to be*

$$[\vec{t}] = \{\vec{s} \mid \vec{s} \sim \vec{t}\},$$

i.e., the set of all positive integer values vectors differing from $\vec{t}$ by a symmetry. Define the n-th equivalence class of $\vec{t}$

$$[\vec{t}]_n = \{\vec{s} \mid \vec{s} \sim \vec{t} \quad \text{and all} \quad s_i \le n\}$$

(we may omit the n subscript if n is understood).

Let n be fixed, and set

$$\mathrm{E_{symm}}(w; \vec{t}) = \mathrm{E_{symm}}(w; \vec{t})_n = \sum_{\vec{s} \in [\vec{t}]_n} P(w; \vec{s})$$

$$= n(n-1)\cdots(n-v+1)P(w; \vec{t}),$$

where v is the number of distinct elements of $\vec{t}$. A symmetric walk sum is just the sum of certain $\mathrm{E_{symm}}(w, \vec{t})$'s, and we can write

$$\mathrm{WalkSum}(\mathcal{W}, k, n) = \sum_{(w; [\vec{t}]) \in \mathcal{W}(k,n)} \mathrm{E_{symm}}(w; \vec{t}),$$

where the summation over $(w; [\vec{t}]) \in \mathcal{W}(k, n)$ means that we sum over one $\vec{t}$ in each equivalence class and over all w.

Each $\vec{t} = (t_0, \ldots, t_k)$ has an $\vec{s} \sim \vec{t}$ where the size of $\vec{s}$'s entries are at most $k+1$. So for each k there are a finite number of equivalence classes, $\mathcal{W}(k)$, of $(w; \vec{t})$ such that w is of length k and $\vec{t}$ is of some finite size. We refer to $\mathcal{W}(k)$ as the set of *potential walk classes* of length k (or *potential closed walk classes* when we restrict to those $\vec{t}$'s with $t_k = t_0$). So if $\mathcal{W}$ is size invariant we may write

$$\text{WalkSum}(\mathcal{W}, k, n) = \sum_{(w;[\vec{t}]) \in \mathcal{W}(k)} \text{E}_{\text{symm}}(w; \vec{t})_n,$$

where the right-hand-side summation has a fixed, finite number of summands independent of n (for fixed k).

Our next step is to comment about $\text{E}_{\text{symm}}(w, \vec{t})_n$. Notice that if the conditions $\sigma_i(t_{i-1}) = t_i$ involve determining a_j values of π_j, then

$$(16) \qquad P(w; \vec{t}) = \prod_{i=1}^{d/2} \frac{1}{n(n-1)\cdots(n-a_j+1)} = \prod_{i=1}^{d/2} \frac{(n-a_j)!}{n!}.$$

Let $e = a_1 + \cdots + a_{d/2}$. We have

$$\text{E}_{\text{symm}}(w, \vec{t})_n = n(n-1)\ldots(n-v+1) \prod_{i=1}^{d/2} \frac{1}{n(n-1)\cdots(n-a_j+1)}.$$

Notice that in the power series expansions about $x = 0$ of

$$(1-x)(1-2x)\ldots(1-mx) \quad \text{and} \quad (1-x)^{-1}(1-2x)^{-1}\ldots(1-mx)^{-1},$$

the x^i coefficient is a polynomial (of degree at most $2i$) in m. It follows that there exist polynomials $p_0, p_1, \ldots$ in the variables $a_1, \ldots, a_{d/2}, v$ such that

$$(17) \qquad \text{E}_{\text{symm}}(w, \vec{t})_n = n^{v-e} \sum_{i=0}^{\infty} n^{-i} p_i(a_1, \ldots, a_{d/2}, v)$$

for n sufficiently large.

DEFINITION 5.4. *We define the* expansion polynomials,

$$p_i = p_i(a_1, \ldots, a_{d/2}, v),$$

to be the polynomials that give the expansion in equation (17). Throughout the paper, a_i denotes the number of π_i values determined by the relevant structure (in this case the potential walk, $(w, \vec{t})$).

We remark that since $e = a_1 + \cdots + a_{d/2}$, we have that v is determined from the a_i if $v - e$ is fixed and known; in such situations, the expansion polynomials may be regarded as functions of the a_i alone.

THEOREM 5.5. *For any $w, \vec{t}$ and any integer $q \geq 0$ we have*

$$(18) \qquad \text{E}_{\text{symm}}(w, \vec{t})_n = n^{v-e}\left(p_0 + \frac{p_1}{n} + \cdots + \frac{p_q}{n^q} + \frac{\text{error}}{n^{q+1}}\right)$$

where

$$|\text{error}| \leq \exp\bigl((q+1)k/(n-k)\bigr)\, k^{2q+2},$$

and the p_i are the expansion polynomials.

The proof is contained between Lemma 2.7 and Corollary 2.10 of [**Fri91**], although the proof there has a minor error. We will correct it here and review the entire proof, since we will later need variants of this theorem for $\mathcal{H}_{n,d}, \mathcal{I}_{n,d}, \mathcal{J}_{n,d}$.

PROOF. If
$$g(x) = (1-b_1 x)\cdots(1-b_r x)(1-c_1 x)^{-1}\cdots(1-c_s x)^{-1}, \tag{19}$$
where the b_i and c_j are positive constants, then g's i-th derivative satisfies the bound
$$|g^{(i)}(x)|/i! \le (1 - x c_{\max})^{-i} \left(\sum b_j + \sum c_j\right)^i$$
where $c_{\max}$ is the maximum of the c_j (by equation (6) in [**Fri91**] on page 339). This estimate, using Taylor's theorem, expanding in $x = 1/n$ about $x = 0$, gives the error term for Theorem 5.5 of
$$(1 - \zeta c_{\max})^{-q-1} \left(\sum b_j + \sum c_j\right)^{q+1},$$
for some $\zeta \in [0, 1/n]$. Since
$$-\log(1 - \zeta c_{\max}) = (\zeta c_{\max}) + (\zeta c_{\max})^2/2 + \cdots \le \zeta c_{\max}/(1 - \zeta c_{\max})$$
$$\le (k/n)/(1 - (k/n)) = k/(n-k),$$
we conclude that the error term is at most
$$\exp\bigl((q+1)k/(n-k)\bigr) \left(\sum b_j + \sum c_j\right)^{q+1}. \tag{20}$$
In the case of Theorem 5.5, the $\sum b_j$ represents the sum of $0, 1, \ldots, v-1$, which is at most $\binom{k}{2}$, and the $\sum c_j$ represents the sum over j of all sums of $0, 1, \ldots, a_j - 1$, which is at most $\binom{k}{2}$. Since $2\binom{k}{2} \le k^2$, we get an error term at most $\exp\bigl((q+1)k/(n-k)\bigr) k^{2q+2}$. (See [**Fri91**] for more details.) □

DEFINITION 5.6. *Given a pair, $(w, \vec{t})$, as above, its order is $e - v$, with e, v as above.*

LEMMA 5.7. *Given a word, w, of length k over Π, we have*
$$\sum_{\vec{t} \text{ such that } (w,\vec{t}) \text{ is of order } \ge r} P(w, \vec{t}) \le n \binom{k}{r+1} \left(\frac{k}{n-k}\right)^{r+1},$$
which for $k \le n/2$ is at most $ck^{2r+2} n^{-r}$ for some constant c depending only on r.

PROOF. This can be found in [**Fri91**], second displayed equation and discussion before on page 352; this is the same idea used in the $r = 1$ case proven in [**BS87**]. (The extra factor of n appears here but not in [**Fri91**], since we do not fix the initial vertex of the walk. Also note that the "order," used here, is one less than the "number of coincidences," used in [**Fri91**].) For the ease of reading, we shall discuss these ideas in our notation.

We may evaluate $P(w, \vec{t})$ by considering the steps of the walk $w = \sigma_1 \ldots \sigma_k$ one by one; i.e., we fix a $v_0 \in \{1, \ldots, n\}$ and consider the random walk $v_1 = \sigma_1(v_0)$, $v_2 = \sigma_2(v_1)$, etc. We inductively consider the probability that $(v_0, \ldots, v_s)$ is equivalent to $(t_0, \ldots, t_s)$ for $s = 1, \ldots, k$. Assuming $(v_0, \ldots, v_s)$ is equivalent to $(t_0, \ldots, t_s)$, and assuming we wish to have $(v_0, \ldots, v_{s+1})$ equivalent to $(t_0, \ldots, t_{s+1})$, one can divide the choice and outcome of the random variable $v_{s+1} = \sigma_{s+1}(v_s)$ into three cases: (1) v_{s+1} has already been determined by previous information (i.e., the value $\sigma_{s+1}(v_s)$ has been determined in a previous step); (2) v_{s+1} has not been determined and (2a) v_{s+1} must occur as one of the $v_0, \ldots, v_s$ (i.e., when t_{s+1} occurs previously in $t_0, \ldots, t_s$), or (2b) v_{s+1} must must be different from $v_0, \ldots, v_s$ (i.e., when t_{s+1}

does not occur previously). We call case (1) a forced choice, case (2) a free choice, with case (2a) a coincidence, and case (2b) a generic choice.

For example, if $w = \pi_1 \pi_1 \pi_2 \pi_3$ and $\vec{t} = (1,1,1,2,1)$, then v_1 is a coincidence (since t_1 occurs previously as t_0 and $\pi_1(v_0)$ has not been determined), v_2 is a forced choice (since $\pi_1(v_0) = v_0 = v_1$ has been determined), v_3 is a generic choice (t_3 does not occur in t_0, t_1, t_2), and v_4 is a coincidence.

A choice $v_{s+1} = \sigma_{s+1}(v_s)$ is deterministic if it is a forced choice; a choice occurs with probability $1 - O(s/n)$ if it is a generic choice, and with probability $1/n + O(s/n^2)$ if it is a coincidence. We see that a word of order t has exactly $t+1$ coincidences. If follows that each word of order at least r and length k has some $r+1$ positions of k that can be marked as coincidences. Fixing a marking, each coincidence of the marking occurs with probability at most $k/(n-k)$ (since v_{s+1} must assume one of the at most $s+1 \leq k$ values $v_0, \ldots, v_s$, and at most k values of the permutation σ_{s+1} could have been determined up to that point). Finally v_0 can be chosen in n ways. We conclude the lemma. □

LEMMA 5.8. *For any irreducible word, w, over Π, of length k, there are at most*

$$(21) \qquad \sum_{j=0}^{r} \binom{k}{j} k^j \leq ck^{2r}$$

equivalence classes $[\vec{t}]$ whose order with w is $\leq r - 1$.

PROOF. See the third displayed equation of page 352 of [**Fri91**] and the discussion preceding. For completeness and ease of reading, we repeat the argument here.

If w has order $j - 1$, then it has j coincidences occurring in k places. Again, in each coincidence, $v_{s+1} = \sigma_{s+1}(v_s)$, v_{s+1} is being chosen from at most k values. This gives the left-hand-side of equation (21). For the right-hand-side we notice that

$$\sum_{j=0}^{r} \binom{k}{j} k^j \leq \sum_{j=0}^{r} k^{2j} \leq k^{2r}(1 + k^{-2} + k^{-4} + \cdots) \leq k^{2r}(4/3)$$

for $k \geq 2$. □

THEOREM 5.9. *Let $\mathcal{W}$ be SSIIC and let $r \geq 1$. Then for all $k \leq n/2$ we have*

$$\text{WalkSum}(\mathcal{W}, k, n) = f_0(k) + \frac{f_1(k)}{n} + \cdots + \frac{f_{r-1}(k)}{n^{r-1}} + \frac{\text{error}}{n^r},$$

where

$$f_i(k) = \sum_{j=0}^{r-1} \sum_{(w;[\vec{t}]) \text{ order } j, \in \mathcal{W}(k)} p_{i-j}(w; [\vec{t}])$$

(with p_i the expansion polynomials, which can be viewed as a function of $(w; [\vec{t}])$ since $(w; [\vec{t}])$ determines the a_i and v) and where for some c depending only on r,

$$|\text{error}| \leq ck^{4r}(d-1)^k.$$

PROOF. By Lemma 5.7, we introduce an error of at most $ck^{2r+2}n^{-r}$ per word by ignoring potential walks of order at least r. Each word, w, has at most ck^{2r} associated potential walk classes of order at most $r - 1$ (by Lemma 5.8), and truncating the associated asymptotic expansion, as in equation (18), of each associated potential walk class results in an error of at most ck^{2r} (by Theorem 5.5). So each word, w, of length k involved in $\mathcal{W}$ contributes an error of at most ck^{4r}, and there are at most $d(d-1)^{k-1}$ such words (for $k \geq 1$) since $\mathcal{W}$ consists of only irreducible words. □

2. The Loop

Here we analyze walk sums associated with simple loops. This gives some ideas and a lemma to be used in Chapter 8.

2.1. The Singly Traversed Simple Loop. Let w be an irreducible word over Π of length k and $\vec{t}$ a $k+1$ tuple over $\{1, \ldots, n\}$ as in the previous subsection. If $t_0 = t_k$ and the t_i's are otherwise distinct, we say $(w; \vec{t})$ is a *singly traversed simple loop* or *STSL* for short; we define $\mathcal{W}_{\text{STSL}}$ to be the collection of all STSL's. When $\mathcal{E}(w; \vec{t})$ occurs and $(w; \vec{t}) \in \mathcal{W}_{\text{STSL}}$, the closed walk from t_0 following w traces out a "simple loop" once, that begins and ends at t_0, moving through distinct edges and vertices throughout the closed walk.

Clearly $\mathcal{W}_{\text{STSL}}$ is SSIIC, so according to Theorem 5.9 we have an asymptotic expansion in $1/n$ with coefficients $f_i(k)$ for the associated walk sum. We now briefly indicate why the f_i are d-Ramanujan. This is a mildly tedious exercise, covered (in much greater generality) in [**Fri91**]. We quote the main points here.

First, note that there is exactly one equivalence class $[\vec{t}]$ of $\vec{t}$'s that appear in $\mathcal{W}_{\text{STSL}}$. So we may write $\mathrm{E}_{\text{symm}}(w)$ for $\mathrm{E}_{\text{symm}}(w; \vec{t})$ for any $\vec{t}$ of the equivalence class, and we may write $p_i(w)$ for the $p_i(w; \vec{t})$ in Theorem 5.9 or 5.5 (note also that $v = k$ in the notation of Theorem 5.5).

Let for $\sigma, \tau \in \Pi$, let $\mathrm{Irred}_{k,\sigma,\tau}$ denote the irreducible words of length k beginning with σ and ending with τ. For $w \in \Pi^k$ let $a_i(w)$ denote the number of π_i and π_i^{-1} occurring in w. Since an STSL has no forced choices, the definition of a_i here agrees with that in Definition 5.4.

LEMMA 5.10. *Let $p = p(a_1, \ldots, a_{d/2}, k)$ be a polynomial. For every σ, τ there are polynomials Q_1, Q_2, Q_3 of k of degree at most the degree of p such that*

$$\sum_{w \in \mathrm{Irred}_{k,\sigma,\tau}} p\big(a_1(w), \ldots, a_{d/2}(w), k\big) = (d-1)^k Q_1(k) + (-1)^k Q_2(k) + Q_3(k).$$

PROOF. This is immediate from Lemma 2.11 of [**Fri91**] or Corollary 2.12 (note that our d is $2d$ in [**Fri91**]). □

Note that the formula for the above lemma comes about from the fact that $d-1, -1, 1$ are the eigenvalues of G_{Irred} where G is the graph with one vertex and $d/2$ whole-loops (with G_{Irred} defined as in Section 6).

COROLLARY 5.11. *We have each $f_i(k)$ as in Theorem 5.9 for $\mathcal{W} = \mathcal{W}_{\text{STSL}}$ is d-Ramanujan.*

PROOF. Recall that an $(w, \vec{t}) \in \mathcal{W}_{\text{STSL}}$ is of order 0, and there is only one equivalence class of $\vec{t}$ for STSL's. So we have

$$f_i(k) = \sum_{\sigma \neq \tau^{-1}} \sum_{\mathrm{Irred}_{k,\sigma,\tau}} p_i\big(a_1(w), \ldots, a_{d/2}(w), k\big),$$

where the p_i are the expansion polynomials. The result now follows from Lemma 5.10, with error term bounded by a polynomial in k.

2.2. Simple Loops.

Consider any irreducible $(w, \vec{t})$ that traces out a simple loop, i.e., the vertices and edges visited form one closed walk, but now we don't require that the loop is traversed only once. Corresponding to this geometric picture of a simple loop we can form the associated walk collection of *simple loop closed walks*; such closed walks, being irreducible, must traverse the loop traced out some number of times.

So let $\mathcal{W}_{\mathrm{SL}}$ be the set of $(w, \vec{t})$ pairs with

(1) $w = \sigma_1 \cdots \sigma_k$ irreducible,
(2) $\sigma_1 \neq \sigma_k^{-1}$,
(3) $t_0, \ldots, t_{r-1}$ distinct for some r dividing s,
(4) $t_{i+r} = t_i$ for $0 \leq i \leq k - r$, and
(5) $w = u^s$ for some word u with $rs = k$.

$\mathcal{W}_{\mathrm{SL}}$ is the walk collection of simple loop walks.

Clearly we have

(22) $$\mathrm{WalkSum}(\mathcal{W}_{\mathrm{SL}}, k, n) = \sum_{s|k} \mathrm{WalkSum}(\mathcal{W}_{\mathrm{STSL}}, s, n).$$

We easily conclude the following theorem.

THEOREM 5.12. *The $f_i(k)$ corresponding to $\mathcal{W}_{\mathrm{SL}}$ are d-Ramanujan, and have the same principal term as the $f_i(k)$ corresponding to $\mathcal{W}_{\mathrm{STSL}}$.*

PROOF. Consider the $f_i(k)$ corresponding to $\mathcal{W}_{\mathrm{STSL}}$. By Corollary 5.11, all the f_i are d-Ramanujan. By equation (22), it suffices to show that

$$\widetilde{f}_i(k) = \sum_{s|k} f_i(s)$$

are also d-Ramanujan. Fixing an i and setting $f_i(k) = p(k)(d-1)^k + r(k)$ as the decomposition of f_i into principal and error terms, we see

$$\widetilde{f}_i(k) = p(k)(d-1)^k + \widetilde{r}(k),$$

where

$$|\widetilde{r}(k)| \leq \left| \sum_{s \leq k/2} p(s)(d-1)^s \right| + \sum_{s \leq k} cs^c (d-1)^{s/2}$$

for some c as in Definition 2.1. It is clear that the right-hand-side above is bounded by $(d-1)^{k/2}$ times a polynomial in k. □

3. Forms, Types, and New Types

In this subsection we will classify potential walks, $(w; \vec{t})$, or more generally potential walk classes, according to some characteristics of the subgraph that the walk traces out.

DEFINITION 5.13. *A form, Γ, is an oriented, Π-labelled graph, $G_\Gamma = (V_\Gamma, E_\Gamma)$, with edges and vertices numbered. A specialization of a form, Γ, is an injection $\iota \colon V_\Gamma \to \{1, \ldots, n\}$.*

With each potential walk, $(w; \vec{t})$, we can associate a form, $\Gamma = \Gamma(w; \vec{t})$, with a specialization, ι, as follows. (The form, Γ, is not unique, but is unique up to unique isomorphism, as described below; the specialization is unique given the form.)

(1) Set $V_\Gamma = \{v_1, \ldots, v_r\}$ to be any numbered (v_i numbered i) set of size r, where $r = |V_\Gamma|$ is the number of distinct elements among the t_i (where $\vec{t} = (t_0, \ldots, t_k)$)
(2) $\iota(v_i)$ is the i-th distinct element of the sequence $t_0, t_1, \ldots, t_k$,
(3) Set $E_\Gamma = \{e_1, \ldots, e_m\}$ to be any numbered set of size m (e_i numbered i), where $m = |E_\Gamma|$ is the number of distinct triples $\{(\sigma_i, t_{i-1}, t_i)\}_{i=1,2,\ldots,k}$, where we identify a triple $(\sigma, s, t,)$ with (σ^{-1}, t, s),
(4) if (σ_j, t_{j-1}, t_j) is the r-th distinct tuple in $\{(\sigma_i, t_{i-1}, t_i)\}_{i=1,2,\ldots,k}$ (with the previous identification), then $s(e_r) = (\sigma_j, \iota^{-1}(t_{j-1}), \iota^{-1}(t_j))$ defines the structural map (see Definition 3.1) of G_Γ.

(see the example in Figure 1 explained at the beginning of this section).

In other words, the form is the subgraph traced out by $(w, \vec{t})$, with some additional information (we remember the order in which the vertices and edges are visited, and the direction each edge is first traversed). We say that forms Γ_1 and Γ_2 are *isomorphic* if they are isomorphic as oriented, numbered, Π-labelled graphs. Because of the numbering, there is at most one isomorphism between any two forms (or a form and itself). We say that $(w, \vec{t})$ is *of form* Γ or *associated to* Γ, written $(w, \vec{t}) \in \Gamma$, if one (or any) of the forms associated to $(w, \vec{t})$ is isomorphic to Γ. Given $(w, \vec{t})$, there is always an associated form, Γ, with $V_\Gamma = \{t_0, \ldots, t_k\}$ and associated specialization, ι, being the identity; however, we usually view V_Γ as any numbered set of the right size, since all of our random graph models are symmetric.

If $(w, \vec{t}) \in \Gamma$, then define
$$\mathrm{E}\,[\Gamma]_n = \mathrm{E}_{\mathrm{symm}}(w, \vec{t})_n$$
which depends only on Γ, and not on the particular $(w, \vec{t})$ to which Γ is associated; indeed,

(23) $$\mathrm{E}\,[\Gamma]_n = \frac{n!}{(n-v)!} \prod_{i=1}^{d/2} \frac{(n-a_i)!}{n!},$$

where $v = |V_\Gamma|$, and a_i is the number of edges in Γ labelled with π_i and π_i^{-1} (this is exactly the $a_i(w)$ of equation (16) of any word, w, associated to Γ).

Hence, if $\mathcal{W}$ is symmetric and size invariant, we may write

(24) $$\mathrm{WalkSum}(\mathcal{W}, k, n) = \sum_\Gamma W_\Gamma(\mathcal{W}, k) \mathrm{E}\,[\Gamma]_n,$$

where $W_\Gamma(\mathcal{W}, k)$ is the number of potential walk classes in $\mathcal{W}(k)$ associated to Γ, and we sum over one Γ in each isomorphism class of forms.

DEFINITION 5.14. *A legal walk in a form, Γ, is a walk starting in v_1 that visits all the vertices of G_Γ in order (of their numbering), all the edges in order, and any edge is first traversed in the direction of its orientation. Each legal walk of length k generates a walk class in the natural way.*

The following easy proposition is worth stating formally; it follows from the definitions.

PROPOSITION 5.15. *$W_\Gamma(\mathcal{W}, k)$ is the number of legal walks on Γ of length k.*

DEFINITION 5.16. *The order of a form, Γ, is $\mathrm{ord}(\Gamma) = |E_\Gamma| - |V_\Gamma|$, subject to the Definition 4.6 convention that any self-loop, whole-loop or half-loop, counts as* a single edge. *(The order of a form equals the order of any potential walk to which it is associated.)*

The "form" allows us to group together potential walks that determine the same information on the graph; this can facilitate the task of studying a walk sum. A further tool is the grouping of forms together by their "type," which we now briefly motivate in rough terms to be made precise. A type arises from a form with all its maximal beaded paths collapsed to edges (by the appropriate suppression). Each collapsed edge inherits an edge length and Π^+ labelling from the "form," but in the type we forget this data. As long as we collect forms of a given type by the first and last letters of all the collapsed edge labellings, we can apply Lemma 5.10 at each collapsed edge to study sums of the expansion polynomials in the a_i's (and v) over forms in such collections.

Before defining a "type," we recall that for a form, $\Gamma = \Gamma(w; \vec{t})$, with specialization, ι, the potential walk $(w; \vec{t})$ pulls back under ι^{-1} to a walk on G_Γ; we remark that the vertex and edge numberings of Γ serve to remember in which order the vertices and edges were traversed in the walk.

DEFINITION 5.17. *A* type, T, *is a connected, oriented graph* $G_T = (V_T, E_T)$, *with vertex and edge numberings such that all vertices except possibly the first one are of degree at least 3. A* labelling *of a type means a Π^+-labelling (recall Π^+ is the set of words over Π of length at least 1).*

To each form, Γ, we associate a type, $T = T(\Gamma)$, as follows. Let W be the set of beads of G_Γ numbered greater than 1. We claim W cannot contain a cycle, for then this cycle would be disconnected from the vertex numbered 1. So we may form the suppression of W in G_Γ. This suppression inherits a vertex and edge numbering from G_Γ (ordering a vertex before another in the suppression if it is numbered less in G_Γ, and ordering edges in the suppression by any associated edge in E_Γ). Of course, the Π-labelling of Γ gives rise to a Π^+-labelling of $G_{T(\Gamma)}$.

In other words, the beads of a form are "less important" features, and the type is just the form with these "less important" features suppressed.

A Π^+-labelled type uniquely determines a form, and vice versa. We wish to group together forms corresponding to one type (inducing different Π^+-labellings), the prototypical example being STSL's or SL's discussed earlier in this section. To do this it will be helpful to remember a small part of the labelling, namely the starting and ending letter of each Π^+-label.

DEFINITION 5.18. *A* lettering *of a type, T, is the assignment to each directed edge a starting letter in Π and an ending letter in Π (such that opposite directed edges are lettered with the starting letter of one being the inverse of the ending letter of the other). Given a form, Γ, with $T = T(\Gamma)$, or equivalently a Π^+-labelling of T, the* associated lettering *assigns to each edge the starting and ending letter of the Π^+-label assigned to it in its orientation.*

It turns out that the notion of a type is too coarse to attack the Alon conjecture (but sufficient for the results of [**Fri91**]). The problem is that some types, when their edges take on certain Π^+-labellings, contain supercritical tangles. When such tangles occur (which they can when their order is roughly $O(\sqrt{d})$), we must distinguish where there tangles occur. This is what a "new type" does, where the type edges are partitioned into "long" and "fixed" (in length) edges, and where supercritical tangles lie on type edges that are "fixed." We then modify our walk sums to be "selective" (see Chapter 6), a notion which requires us to know where these tangles occur in the type.

DEFINITION 5.19. *A B-new type is a collection,* $\widetilde{T} = (T; E_{\text{long}}, E_{\text{fixed}}; \vec{k}^{\text{fixed}})$, *of (1) a lettered type, T, (2) a partition of E_T into two sets, $E_{\text{long}}, E_{\text{fixed}}$, (3) for each $e_i \in E_{\text{fixed}}$ an edge length, k_i^{fixed}, with $0 < k_i^{\text{fixed}} < B$, and (4) a Π^+-labelling of E_{fixed} with each $e_i \in E_{\text{fixed}}$ labelled with a word of length k_i^{fixed}. A Π^+-labelling of T (or, equivalently, a form, Γ), is said to be of B-new type $\widetilde{T}$ if each E_{long} label is of length at least B, and each label corresponding to $e_i \in E_{\text{fixed}}$ is of length k_i^{fixed} and agrees with the label specified by $\widetilde{T}$. $\widetilde{T}$ is said to be* based on T.

THEOREM 5.20. *For each $r > 0$ there are finitely many types (up to isomorphism) of order at most r. For each type, T, and each $B > 0$ there are finitely many B-new types based on T.*

PROOF. The first statement is just Lemma 2.3 of [**Fri91**], except that "coincidence" is used instead of "order" (and the coincidence is the order plus one). The second statement is clear since there are finitely many (1) letterings, (2) partitions of E_T, (3) choices of k_i^{fixed} with $e_i \in E_{\text{fixed}}$, and (4) labellings of each E_{fixed} edge, e_i, with a length $k_i^{\text{fixed}} < B$. □

4. Motivation of Types and New Types

So far we have defined walk sums; we have seen that symmetric, size invariant walk sums can be organized into forms, by equation (24); we have seen that forms can be grouped by type and new type. In this section we briefly explain how and why we use types and new types.

Fix a symmetric, size invariant walk collection, $\mathcal{W}$. Organizing forms by type, we may write equation (24) as

$$\text{WalkSum}(\mathcal{W}, k, n) = \sum_T \sum_{\Gamma \in T} W_\Gamma(\mathcal{W}, k) \text{E}\left[\Gamma\right]_n,$$

where $\Gamma \in T$ means the form Γ is of type T, and the summation in T ranges over all types. Typically we need only sum over T of at most some order, so the sum in T will effectively be a finite sum.

So fix a T and define $\text{E}\left[T\right]_{n,k}$ to mean

$$\text{E}\left[T\right]_{n,k} = \sum_{\Gamma \in T} W_\Gamma(\mathcal{W}, k) \text{E}\left[\Gamma\right]_n.$$

Let T's edges be $E_T = \{e_1, \ldots, e_b\}$. For a vector, $\vec{k} = (k_1, \ldots, k_b)$, let $T(\vec{k})$ denote the set of forms, Γ, of type T, such that for all i the length of the beaded path in Γ corresponding to the edge e_i in T has length k_i. For each $e_i \in E_T$ fix an integer $m_i \geq 1$. Let $W_\Gamma(\vec{m}) = W_\Gamma(m_1, \ldots, m_b)$ denote the number of legal $\mathcal{W}$ walks in Γ that traverse edge e_i exactly m_i times. Clearly

$$W_\Gamma(\vec{m}) = W_T(\vec{m}),$$

i.e., $W_\Gamma(\vec{m})$ depends only on $\vec{m}$ and the type, T, of Γ. This allows us to write

$$\text{E}\left[T\right]_{n,k} = \sum_{\vec{m}} W_T(\vec{m}) \sum_{\vec{k} \cdot \vec{m} = k} \sum_{\Gamma \in T(\vec{k})} \text{E}\left[\Gamma\right]_n.$$

Since each $\mathrm{E}\,[\Gamma]_n$ has a $1/n$ expansion, by adding expansions we get an asymptotic expansion
$$\sum_{\Gamma \in T(\vec{k}\,)} \mathrm{E}\,[\Gamma]_n = n^{-\mathrm{ord}(T)}\bigl(q_0(\vec{k}\,) + q_1(\vec{k}\,)/n + \cdots\bigr),$$
where we ignore error terms in this subsection; of course,
$$q_i(\vec{k}\,) = \sum_{\Gamma \in T(\vec{k}\,)} p_i\bigl(a_1(\Gamma), \ldots, a_{d/2}(\Gamma)\bigr).$$

Perhaps the main technical point of Chapter 8 (see Theorem 8.5 here, and Lemma 2.14 in [**Fri91**]) is that the $P_{\vec{m},i}$ defined by
$$P_{\vec{m},i}(k) = \sum_{\vec{k}\cdot\vec{m}=k} q_i(\vec{k}\,)$$
are d-Ramanujan, and roughly speaking $P_{\vec{m},i}$'s principal part and error term decay at most like $(d-1)^{m/2}$ where $m = m_1 + \cdots + m_k$ (actually, the principal part is shown in Lemma 2.14 in [**Fri91**] to decay at most like $(d-1)^{-m}$). The way this is done is very roughly as follows. First, we fix a lettering of T and apply Lemma 5.10 to each edge, e_i, of T, whose length is set to k_i. This shows that the $q_i(\vec{k}\,)$ is a sum of d-Ramanujan functions whose arguments are all sums of a subset of the k_i. As an example, consider a function, $g(k_1 + \cdots + k_b)$ with g being d-Ramanujan. The maximum value of $k_1 + \cdots + k_b$, given $\vec{k}\cdot\vec{m} = k$ with fixed $\vec{m}$ and k, is $k - m + b$, which is achieved when and only when $k_i = 1$ whenever $m_i \geq 2$ (assuming at least one $m_i = 1$; otherwise $g(k_1 + \cdots + k_b)$ is bounded by $ck^c(d-1)^{k/2}$ since $k_1 + \cdots + k_b \leq k/2$). This $k - m + b$ is where the decay exponential in m comes from (see Section 8 and/or Section 2 in [**Fri91**] for details).

It follows that there is an asymptotic expansion
$$\mathrm{E}\,[T]_{n,k} = n^{-\mathrm{ord}(T)}\bigl(Q_0(k) + Q_1(k)/n + \cdots\bigr),$$
where
(25) $$Q_i(k) = \sum_{\vec{m}} W_T(\vec{m}\,)P_{\vec{m},i}(k).$$

Since the error term of the $P_{\vec{m},i}$ decay like $(d-1)^{-m/2}$ (roughly speaking), the above sum for $Q_i(k)$ turns out to be d-Ramanujan provided that
(26) $$|W_T(\vec{m}\,)| \leq c(d-1-\epsilon)^{m/2}$$
for some $\epsilon > 0$.

If all Q_i were d-Ramanujan for all T, then we'd have a fairly simple proof of the Alon conjecture. The results of [**Fri91**] are based on the fact that equation (26) for all types up to a certain order (of roughly $O(\sqrt{d})$); past order $O(\sqrt{d})$, equation (26) generally fails to hold. Furthermore, by Theorem 2.12 we see for certain that at least one Q_i fails to be d-Ramanujan for an i bounded by $O(\log d\sqrt{d})$.

The new approach in this paper is as follows. Consider a constant positive integer B, and consider
$$Q_i(k) = \sum_{\vec{m}} W_T(\vec{m}\,) \sum_{\vec{k}\cdot\vec{m}=k} q_i(\vec{k}\,).$$

We can divide this sum by fixing some of the k_i's at fixed values less than B (call these $k_{t+1}, \ldots, k_b$), and then summing over the remaining k_i's (namely $k_1, \ldots, k_t$)

subject to the remaining k_i's being at least B (or "long"). This is where a "new type" comes from. Next we fix a constant $S < B$ and define a "selective trace" to be the sum of irreducible closed walks of a given length that have no subpath of length S tracing out a graph of λ_{Irred} at least $\sqrt{d-1}$. If

$$M_1 = m_1 + \ldots + m_t, \quad M_2 = m_{t+1} + \ldots + m_b,$$

then the corresponding "selective" version of $W_T(\vec{m})$ (that depends on the new type, i.e., on knowing the particular k_i that are fixed and their values) is bounded by roughly $(d-1-\epsilon)^{(BM_1+M_2)/2}$ (for appropriately large S and B). But consider the $P_{\vec{m},i}$ for the "new type," i.e., the sum

$$\sum_{\substack{\vec{k}\cdot\vec{m}=k \\ k_1,\ldots,k_t \geq B}} q_i(\vec{k})$$

with $k_{t+1}, \ldots, k_b$ at their fixed values; it turns out they decay like $(d-1)^{-(BM_1+M_2)/2}$. Thus, after summing over all new types, the corresponding selective analogues of the $Q_i(k)$ are d-Ramanujan.

Unfortunately, a selective trace does not generally equal the original trace unless the graph in question is free of certain tangles. Still, in Chapters 9 and later we show how the asymptotic expansions with d-Ramanujan coefficients of selective traces can be used to control a random graph's eigenvalues.

CHAPTER 6

The Selective Trace

In this section we define a *selective trace*, and discuss some of its properties.

1. The General Selective Trace

Fix a graph, $G = (V, E)$, coming from $\mathcal{G}_{n,d}$, so that $V = \{1, \ldots, n\}$ and G is Π-labelled.

By a *path*[1] *of length* k in G we shall mean a vertex, $v \in V$, and a word of length k, $w = \sigma_1 \ldots \sigma_k$, over Π (i.e., with each $\sigma_i \in \Pi$). Such a path determines a subgraph in G of those vertices and edges traversed. We say a path *traverses* a tangle, ψ, if the subgraph traversed by the path contains the tangle, ψ.

DEFINITION 6.1. *Let* $\Psi = \{\psi_1, \ldots\}$ *be a (finite or infinite) collection of tangles. For positive integer,* S, *the set of* (S, Ψ)*-selective closed walks (respectively, walks) are those irreducible closed walks (respectively, walks) that have no subpath of length at most* S *that traverses a tangle in* Ψ. *The* k*-th irreducible* (S, Ψ)*-selective trace of* G, IrSelTr$_{S,\Psi}(G; k)$, *is the number of* (S, Ψ)*-selective closed walks of length* k.

Intuitively, the selective trace modifies the standard irreducible trace on those graphs that have a tangle in Ψ, and avoids those closed walks that in some short part trace out such a tangle.

2. A Lemma on Selective Walks

What is the point of the selective trace? We can answer this question in two ways. First, since hypercritical tangles give large eigenvalues, any trace with an arbitrarily long asymptotic expansion in $1/n$ with d-Ramanujan coefficients must avoid hypercritical tangles (according to Theorems 3.13 and 4.2); a trace must be selective or its asymptotic expansion coefficients will not all be d-Ramanujan. Second, there is a crucial technical theorem, Theorem 6.6, about counting irreducible contributions to a selective trace. This lemma makes certain infinite sums converge for the selective trace that would have to diverge for the standard trace— for example, the infinite sum involving $W(T; \vec{m})$ and $P_{i,T,\vec{m}}$ just above the middle of page 351 in [**Fri91**], for types of order $> d$; for the same reason, this crucial theorem makes the $1/n$ expansion for a selective trace have d-Ramanujan coefficients when they don't for a trace that is not selective— indeed, the $(2d-1)^{k/2}$ bound in equation (24) of [**Fri91**] depends critically on $2i + 2 \leq \sqrt{2d-1}$, and this equation corresponds to the error in the n^{-i} term in the expansion of the expected value

---

[1]By a *path* one often means a sequence of vertices. In case there are multiple edges in the graph, one needs to note also which edge is traversed. Finally, in the case of whole-loops in an undirected graph, one needs to remember in which "direction" each whole-loop is being traversed. In the present situation, all the above information is contained simply in the initial vertex and the permutations, π_i or π_i^{-1}, being taken on each step of the path.

of the irreducible trace (recall that $2d$ in [**Fri91**] corresponds to our d). We shall finish this subsection with the crucial technical theorem, Theorem 6.6, after setting up the necessary terminology.

A *relabelling* of a tangle, ψ, is a tangle, ψ', that differs from ψ only in its edge labels.

DEFINITION 6.2. *A set, Ψ, of tangles is called* closed under pruning *(respectively, relabelling) if $\psi \in \Psi$ implies $\psi' \in \Psi$ for any pruning (respectively, relabelling), ψ', of ψ.*

Note: In the definition above, ψ and ψ' must be $\mathcal{G}_{n,d}$-tangles (or tangles in whatever model is discussed)— a vertex with two self-loops labelled both labelled π_1 is not a $\mathcal{G}_{n,d}$-tangle and is therefore not considered a relabelling of the tangle where the self-loops are labelled π_1 and π_2.

DEFINITION 6.3. *For a positive integer, τ, let $\Psi_{\mathrm{ord}}(\tau)$ be the set of tangles whose order is at least τ. For positive integers $\tau_1 \leq \tau_2$, let $\Psi_{\mathrm{ord}}(\tau_1, \tau_2)$ be the set of all tangles whose order is at least τ_1 and at most τ_2. We also write $\Psi_{\mathrm{ord}}(\tau, \infty)$ for $\Psi_{\mathrm{ord}}(\tau)$.*

Since pruning a tangle does not affect its order, $\Psi_{\mathrm{ord}}(\tau_1, \tau_2)$ is closed under pruning; clearly $\Psi_{\mathrm{ord}}(\tau_1, \tau_2)$ is closed under relabelling.

Consider a form, Γ, of type T, in which T's edges, e_i, have length k_i (as beaded paths arising from Γ). For each $e_i \in E_T$ fix an integer $m_i \geq 1$. Let Ψ be a set of tangles closed under relabelling. Let $W_\Gamma(\vec{m}; S, \Psi)$ be the number of legal closed walks (in particular, beginning at the first vertex) in Γ that traverse each e_i exactly m_i times (in either direction) and that are (S, Ψ)-selective. Since Ψ is closed under relabelling, W_Γ depends only on the length, k_i, of e_i in Γ, not on the particular Π^+ labels of length k_i. So we may write

$$W_\Gamma(\vec{m}; S, \Psi) = W_T(\vec{m}, \vec{k}; S, \Psi).$$

Now given the above setting, call an edge, e_i, of T *long* if $k_i > S$, and *short* otherwise. If a walk contains some $\psi \in \Psi$ in any consecutive S steps, then by possibly pruning these consecutive steps along long edges at the beginning and end, we get a consecutive walk over short edges that contains a pruning of ψ. In particular, if $B > S$ and $\widetilde{T} = (T; E_{\mathrm{long}}, E_{\mathrm{fixed}}; \vec{k}^{\mathrm{fixed}})$ is a B-new type based on T, then W_T depends on only $\widetilde{T}$, $\vec{m}$, S, and Ψ provided that $k_i = k_i^{\mathrm{fixed}}$; hence we may write

$$W_T(\vec{m}, \vec{k}; S, \Psi) = W_{\widetilde{T}}(\vec{m}; S, \Psi).$$

DEFINITION 6.4. *We say that a collection of tangles, Ψ, is r-supercritical if it contains all supercritical tangles of order at most r.*

Next we give two examples of very natural r-supercritical tangle sets.

DEFINITION 6.5. *Let τ_{fund} be the smallest order of a supercritical tangle, and let $\Psi_{\mathrm{fund}} = \Psi_{\mathrm{ord}}(\tau_{\mathrm{fund}})$. Let Ψ_{eig} the be set of all supercritical tangles.*

Ψ_{fund} and Ψ_{eig} are clearly r-supercritical for any r; $\Psi_{\mathrm{ord}}(\tau_{\mathrm{fund}}, r)$ and $\Psi_{\mathrm{eig}}[r] = \Psi_{\mathrm{eig}} \cap \Psi_{\mathrm{ord}}(\tau_{\mathrm{fund}}, r)$ are also clearly r-supercritical. We arrive at our crucial technical theorem, that is the key to the selective trace.

2. A LEMMA ON SELECTIVE WALKS

THEOREM 6.6. *Let T by any type, with specified edge set partition $E_{\text{long}}, E_{\text{fixed}}$, and a Π-lettering specified. Let the edges be indexed so that*

$$E_{\text{long}} = \{e_1, \ldots, e_t\}, \qquad E_{\text{fixed}} = \{e_{t+1}, \ldots, e_b\}.$$

Then there is a c, an $\epsilon > 0$, and an S_0 such that the following is true for all $S \geq S_0$. Let Ψ be a set of tangles containing all supercritical tangles included in a form of type T; e.g., by Lemma 4.8 we may take Ψ to be any r-supercritical set for $r = \text{ord}(T)$. Let

$$W_{\widetilde{T},S}(M_1, M_2) = \sum_{\substack{m_1 + \ldots + m_t = M_1 \\ m_{t+1} + \cdots + m_b = M_2 \\ m_i \geq 1}} W_{\widetilde{T}}(\vec{m}; S, \Psi),$$

for a B-new type, $\widetilde{T}$, with $B > S$ and with $\widetilde{T}$ having edge set partition $E_{\text{long}}, E_{\text{fixed}}$. Then

$$W_{\widetilde{T},S}(M_1, M_2) \leq c(\sqrt{d-1} - \epsilon)^{(S_0+1)M_1 + M_2} \leq c(\sqrt{d-1} - \epsilon)^{BM_1 + M_2}.$$

PROOF. The general approach we take is based on the following crude estimate. Let $\{f_i(z)\}_{i \in I}$ be a collection of non-negative power series, and let $g(z)$ be a power series that majorizes each $f_i(z)$. Suppose that $g(z_0)$ converges for some $z_0 \in (0,1)$. Then the z^k coefficient of any of the $f_i(z)$'s is bounded by $g(z_0) z_0^{-k}$.

Specifically, we shall show that there is an S_0 such that the following holds. Let $B > S_0$, and let $\widetilde{T}$ be a B-new type subject to the conditions of the theorem. Let G be a VLG whose underlying graph is T, whose E_{fixed} edges take their lengths from $\widetilde{T}$, and whose E_{long} edges all have length B. Let $f(z) = \sum c_k z^k$, where c_k is the number of (S_0, Ψ)-selective irreducible walks of length k in G. We shall show that there is a $z_0 > (d-1)^{-1/2}$ and a $g = g(z)$ such that (1) g majorizes f, (2) g depends only on T, and (3) $g(z_0)$ converges. In that case

$$W_{\widetilde{T},S}(M_1, M_2) \leq g(z_0) z_0^{-BM_1 - M_2},$$

which completes the theorem.

Let $\mathcal{G}_{\text{below}}$ be the set of VLG's, H, whose underlying graph is a subgraph of T containing only E_{fixed} edges, with the property that $\lambda_{\text{Irred}}(H) < \sqrt{d-1}$. Let $\mathcal{G}_{\text{extreme}}$ be the set of elements of $\mathcal{G}_{\text{below}}$ that are not majorized by a different member of $\mathcal{G}_{\text{below}}$. We claim that $\mathcal{G}_{\text{extreme}}$ is finite; indeed, if $H_1, H_2, \ldots$ were a distinct sequence of VLG's in $\mathcal{G}_{\text{extreme}}$, then by passing to a subsequence we could assume that for every $e \in E_{\text{fixed}}$ either the length of e in H_i is constant or the length tends to infinity; but then H_1 would majorize all H_i with i sufficiently large.

Let $\mathcal{G}_{\text{extreme}} = \{H_1, \ldots, H_m\}$, and let $h_i(z) = \sum c_{i,k} z^k$, where $c_{i,k}$ is the number of irreducible walks of length k in H_i. These $c_{i,k}$ are given as the number of walks of length $k-1$ in $(H_i)_{\text{Irred}}$; it follows that $h_i(z)$ has radius of convergence greater than $(d-1)^{-1/2}$. Also, if $\widetilde{c}_{i,k} = c_{i,0} + \cdots + c_{i,k}$ is the number of irreducible walks in H_i of length at most k, then

$$\widetilde{h}_i(z) = \sum_k \widetilde{c}_{i,k} z^k = \frac{1}{1-z} h_i(z)$$

also has radius of convergence greater than $(d-1)^{-1/2}$. Let $z_0 > (d-1)^{-1/2}$ be a value at which all $\widetilde{h}_i$ converge.

For any S set
$$\widetilde{h}_i^S(z) = \sum_{k>S} \widetilde{c}_{i,k} z^k,$$
and set
$$\widetilde{h}^S(z) = \sum_{i=1}^m \widetilde{h}_i^S(z) \quad \text{and} \quad h(z) = \sum_{i=1}^m h_i(z).$$

Consider an S_0 sufficiently large so that $\widetilde{h}^{S_0}(z_0) < 1/d$ (later we impose other lower bounds on S_0's value).

Let $B > S_0$. Let G be a VLG whose underlying graph is (the graph underlying) T, and whose E_{long} edges have length at least B. Let d_k (respectively, $\widehat{d}_k$) be the number of irreducible walks in G of length k that are (S_0, Ψ)-selective (respectively, and never traverse an E_{long} edge). Let $f(z), \widehat{f}(z)$ be the generating functions of $d_k, \widehat{d}_k$, respectively.

The functions $f(z), \widehat{f}(z)$ are clearly majorized by the $f(z), \widehat{f}(z)$ in the case where $B = S_0 + 1$ and all E_{long} edges have length $S_0 + 1$. We shall assume this to be the case.

First, we claim that f is majorized by
$$\left(1 - \widehat{f}(z)dz^{S_0+1}\right)^{-1} \widehat{f}(z).$$

This is because any walk in G can be broken into alternating subwalks that remain in E_{fixed} and steps along E_{long} edges; each time an E_{long} edge is taken its length is at least $S_0 + 1$, and there are at most d such E_{long} edges from which to choose after finishing the E_{fixed} walk[2]. To prove the theorem it therefore suffices to show that $\widehat{f}(z_0)dz_0^{S_0+1}$ is less than one for sufficiently large S_0.

Next, we claim that $\widehat{f}(z)$ is majorized by
$$\left(1 - d\widetilde{h}^{S_0}(z)\right)^{-1} h(z).$$

Assuming this, it is clear that for sufficiently large S_0 we have $\widehat{f}(z_0)dz_0^{S_0+1} < 1$ and the theorem is proven. For $k \geq S_0$, let b_k be the number of (S_0, Ψ)-selective irreducible walks, $w = (v_0, e_1, v_1, \ldots, e_j, v_j)$, in G of length k through only E_{fixed} edges such that $w' = (v_0, e_1, v_1, \ldots, e_{j-1}, v_{j-1})$ is of length less than S_0. The walk w' is contained in a subgraph of G that is majorized by one of the H_i. Clearly
$$b_{S_0} \leq \sum_{i=1}^m c_{i,S_0}.$$

When $k > S_0$, then from v_{j-1} the walk has at most d possible directions to take (or at most $d - 1$ possible directions if $j > 1$), and so
$$b_k \leq \sum_{i=1}^m \widetilde{c}_{i,k}.$$

Thus $\sum b_k z^k$ is majorized by $d\widetilde{h}^{S_0}(z)$. But any walk over E_{fixed} edges can be broken into a series of walks of the form w as above, plus a final walk of length less than S_0. The generating function for such walks is clearly majorized by $h(z)$. □

---

[2]Of course, there are at most $d - 1$ such E_{long} edges from which to choose except possibly at the very first step of a walk.

3. Determining τ_{fund} for $\mathcal{G}_{n,d}$

In order to use the selective trace, we must determine τ_{fund}. We begin by doing so for the model $\mathcal{G}_{n,d}$; next subsection we use similar techniques for the models $\mathcal{H}_{n,d}$, $\mathcal{I}_{n,d}$, and $\mathcal{J}_{n,d}$.

More generally, for a given τ, consider the task of finding the tangle, ψ, in $\mathcal{G}_{n,d}$, of order at most τ with $\lambda_{\text{Irred}}(\psi)$ as large as possible. To simplify this task, notice that pruning leaves the order and λ_{Irred} invariant (an irreducible closed walk can never visit a leaf, so pruning a leaf doesn't affect the number of irreducible closed walks); hence we may restrict our search to those ψ's that are completely pruned.

LEMMA 6.7. *Let G be a graph with edge $e = \{u, v\}$ with $u \neq v$. Let G_e be the contraction of G along e, i.e. the graph obtained by discarding e and identifying u with v. Then $\lambda_{\text{Irred}}(G) \leq \lambda_{\text{Irred}}(G_e)$.*

PROOF. Consider an irreducible closed walk, c, about u in G. Then we can associate to this closed walk one in G_e, $\iota(c)$, by discarding all occurrences of e. This association, ι, is an injection, since given a G_e irreducible closed walk about u of the form $\iota(c)$, we can infer when e was taken (since $e = \{u, v\}$ with $u \neq v$) in the G closed walk, giving rise to (at most) a single G closed walk. Since this injection does not increase the length of the closed walks, we conclude that the number of irreducible closed walks about u in G of length $\leq k$ is no more than the number in G_e. Hence the conclusion of the lemma. □

Since edge contraction reduces the number of vertices and of edges by one each, edge contraction leaves the order invariant. So in looking for a λ_{Irred} tangle of a given order, we may always assume the tangle has no edge contraction that leaves it a tangle[3].

We now claim (by Lemma 6.7) that for $\mathcal{G}_{n,d}$ and $\tau \leq (d/2) - 1$, a vertex with $\tau + 1$ whole-loops has the largest λ_{Irred} of all tangles of order τ (recall that each self-loop is counted as one edge, according to Definition 4.6). For this graph we clearly have $\lambda_{\text{Irred}} = 2\tau + 1$; hence τ_{fund} is the smallest integer τ with $2\tau + 1 \geq \sqrt{d-1}$, provided that this τ is $\leq (d/2) - 1$. But we easily verify that this τ,

$$\tau_{\text{fund}} = \lceil (\sqrt{d-1} + 1)/2 \rceil - 1 = \lceil (\sqrt{d-1} - 1)/2 \rceil$$

is indeed at most $(d/2) - 1$ for all $d \geq 4$. We have just established the following theorem.

THEOREM 6.8. *For the model $\mathcal{G}_{n,d}$, we have $\tau_{\text{fund}} = \lceil (\sqrt{d-1} + 1)/2 \rceil - 1$.*

4. Determining τ_{fund} for $\mathcal{H}_{n,d}$, $\mathcal{I}_{n,d}$, and $\mathcal{J}_{n,d}$

For $\mathcal{H}_{n,d}$ we have to remember that tangles can't have self-loops. Thus contractions can only be done along non-multiple edges, and τ_{fund} will not generally be the same for $\mathcal{H}_{n,d}$ and $\mathcal{G}_{n,d}$.

LEMMA 6.9. *Let u, v be vertices of distance two in a graph, G, i.e., there are no edges joining u and v, but there is a w with edges to each of u, v. Let G' be the graph*

---

[3]In $\mathcal{H}_{n,d}$, two vertices joined by between 2 and $d/2$ edges is a tangle (with appropriate Π-labelling), but contracting any edge gives self-loops, which are not feasible in $\mathcal{H}_{n,d}$. Therefore edge contraction can take graphs that can be tangles to graphs that cannot, at least for certain random graph models.

obtained by identifying u and v and deleting one of the edges from w to u (or to v) (so that the order of G' is the same as that of G). Then $\lambda_{\text{Irred}}(G) \leq \lambda_{\text{Irred}}(G')$.

PROOF. Let U be the vertex in G' which is the identification of v and u. Let the edges from u to w be enumerated $e_1, \ldots, e_s$, and those from v to w enumerated $f_1, \ldots, f_t$. The edges from U to w are $g_1, \ldots, g_r$, where $r = s + t - 1$.

First consider the case when $V_G = \{u, v, w\}$, and consider the irreducible closed walks about w (which are necessary of even length). Such a closed walk begins in w and takes two steps, visiting either u or v, in, respectively, $s(s-1)$ or $t(t-1)$ ways. After coming back from a u vertex, another step of length 2 can either (1) visit a u vertex, in $(s-1)^2$ ways, or (2) visit a v vertex, in $t(t-1)$ ways; similarly for coming back from a v vertex. Thus "coming back from a u vertex" and "coming back from a v vertex" forms a Markov chain, and the total number of irreducible closed walks of length k about w is

$$(27) \quad I_1(k) = \begin{bmatrix} s(s-1) & t(t-1) \end{bmatrix} \begin{bmatrix} (s-1)^2 & t(t-1) \\ s(s-1) & (t-1)^2 \end{bmatrix}^{(k-2)/2} \begin{bmatrix} 1 \\ 1 \end{bmatrix}$$

We wish to compare this to the number of irreducible G' closed walks about w, of which there are clearly

$$I_2(k) = r(r-1)^{k-1} = (s+t-1)(s+t-2)^{k-1}.$$

For starters, we see

$$I_2(2) - I_1(2) = 2(s-1)(t-1)$$

which is non-negative, since both $s, t \geq 1$. Now since the maximum row sum in the 2×2 matrix of equation (27) is

$$s^2 + t^2 - 2(s+t) + 1 + \max(s, t),$$

we have

$$I_1(k+2) \leq I_1(k) m_1, \quad \text{where} \quad m_1 = s^2 + t^2 - 2(s+t) + 1 + \max(s, t)$$

for all k. But

$$I_2(k+2) = I_2(k) m_2, \quad \text{where} \quad m_2 = (s+t-2)^2,$$

and

$$m_2 - m_1 = 2st - 2(s+t) + 3 - \max(s,t) = 1 + 2(s-1)(t-1) - \max(s,t),$$

which is positive unless s or t is 1. Thus, provided that $s \geq 2$ and $t \geq 2$, we have

$$\lambda_{\text{Irred}}(G) \leq \sqrt{m_1} < \sqrt{m_2} = \lambda_{\text{Irred}}(G'),$$

and

$$(28) \quad I_1(k) \leq I_1(2) m_1^{(k-2)/2} < I_2(2) m_2^{(k-2)/2} I_2(k) = I_2(k)$$

for all even k. If $t = 1$ we calculate

$$(29) \quad I_1(k) = s(s-1)^{k-1} = I_2(k),$$

and similarly when $s = 1$.

We shall use the above calculation below. We can now assume that V_G has a vertex, x, different from u, v, w.

There is a natural bijection of edges, ι from $E_G \setminus (\{e_i\} \cup \{f_i\})$ to $E_{G'} \setminus \{g_i\}$. Extend ι to a map on all of E_G by defining $\iota(e_i)$ and $\iota(f_i)$ to be a formal symbol

S. For any irreducible G closed walk about x specified by its edges, $c = (c_1, \ldots, c_k)$ with $c_i \in E_G$, we associate a sequence
$$\iota(c) = (\iota(c_1), \ldots, \iota(c_\ell)).$$
We claim that the number of c with a given image $\iota(c)$ is no more than the number of $E_{G'}$ closed walks corresponding to $\iota(c)$ by changing all g_i edges into S's. Indeed, consider a block of consecutive S's in $\iota(c)$, i.e. $\iota(c_a) = \iota(c_{a+1}) = \cdots = \iota(c_b) = S$, and $\iota(c_{a-1}) \ne S$ and $\iota(c_{b+1}) \ne S$; $\iota(c)$ cannot begin or end with an S, since the closed walk begins at x, and so we can assume $a \ge 2$ and $b \le \ell - 1$. By looking at $\iota(c_{a-1})$ and $\iota(c_{b+1})$ we can determine whether or not the S-block begins in u, v, or w, and ends in u, v, or w. If the S-block begins in w and ends in w, then equations (28) and (29) show that there are no fewer G' sequences for the corresponding S-block than G sequences. Next compare those S-blocks that begin in a u and end in a w. The number of such sequences in G is
$$\begin{bmatrix} s & 0 \end{bmatrix} \begin{bmatrix} (s-1)^2 & t(t-1) \\ s(s-1) & (t-1)^2 \end{bmatrix}^{(b-a)/2} \begin{bmatrix} 1 \\ 1 \end{bmatrix},$$
whereas the number in G' is $(s+t-1)(s+t-2)^{b-a}$ (since the non-S edge $\iota(c_{a-1})$ can be followed by any U to w edge in G'); so the G' number is no less than the G number for $b-a = 0$ (since $t \ge 1$), and each time $b-a$ is increased by 2, the former number gets multiplied by an m_2, the latter gets multiplied by no more than m_1, where $m_1 < m_2$, provided that $s \ge 2$ and $t \ge 2$; the $s = 1$ or $t = 1$ case is easily checked to result in equality. The same argument holds for v to w S-blocks. For an S-block starting and ending in u, we wish to compare
$$\begin{bmatrix} s & 0 \end{bmatrix} \begin{bmatrix} (s-1)^2 & t(t-1) \\ s(s-1) & (t-1)^2 \end{bmatrix}^{(b-a-1)/2} \begin{bmatrix} s-1 \\ 0 \end{bmatrix},$$
with $(s+t-1)(s+t-2)^{b-a}$. Again, it suffices to compare when $b-a = 1$, which is immediate, and to check $s = 1$ or $t = 1$ separately. We argue for S-blocks starting in either u or v and ending in either u or v similarly. □

THEOREM 6.10. *For the model $\mathcal{H}_{n,d}$, we have $\tau_{\text{fund}} = \lceil \sqrt{d-1} \rceil - 1$.*

PROOF. As before, we consider a τ and search for those ψ of order at most τ with $\lambda_{\text{Irred}}(\psi)$ as large as possible. By Lemma 6.9, and by contractions (in Lemma 6.7), we may restrict our search to those ψ with two or more edges between every pair of nodes.

First assume that $\tau + 2 \le d/2$. If ψ has two vertices, then ψ has $\tau + 2$ edges joining the two vertices (since there are no self-loops in $\mathcal{H}_{n,d}$). In this case $\lambda_{\text{Irred}}(\psi) = \tau + 1$. We claim that this is as large a λ_{Irred} as possible (again, assuming $\tau + 2 \le d/2$). Indeed, if ψ has $r > 2$ vertices, then the maximum degree of a vertex is $|E|$ minus the edges not involved with that particular vertex, which is at least 2 for each pair of the $r-1$ other vertices. So the maximum degree is at most
$$|E| - \binom{r-1}{2} 2 \le (|V| + \tau) - \binom{r-1}{2} 2 = \tau + r - (r-1)(r-2).$$
Since λ_{Irred} is no greater than the maximum degree minus 1, we have
$$\lambda_{\text{Irred}} \le \tau + r - (r-1)(r-2) - 1 = \tau + 1 - (r-2)^2.$$
It follows that if $r > 2$, λ_{Irred} is strictly less than $\tau + 1$.

To achieve $\lambda_{\text{Irred}}(\psi) = \tau + 1$ with our ψ having two vertices, we required $\tau + 2 \le d/2$. To get $\lambda_{\text{Irred}}(\psi) = \tau + 1$ to equal or exceed $\sqrt{d-1}$, we require $\tau + 1 = \lceil \sqrt{d-1} \rceil$, for which we must have

$$\lceil \sqrt{d-1} \rceil + 1 \le d/2.$$

Since $d/2$ is an integer, this is equivalent to

$$\sqrt{d-1} + 1 \le d/2,$$

which we easily see holds for all even $d > 2$ except $d = 4, 6$.

We conclude that $\tau_{\text{fund}} = \lceil \sqrt{d-1} \rceil - 1$ for even $d \ge 8$. It suffices to analyze the cases $d = 4, 6$.

For each order, τ, and $d = 4, 6$, we must examine those tangles of order τ and determine the largest possible $\lambda_{\text{Irred}}(\psi)$. Let us note that if ψ is a tangle of order -1, then it is a tree and has $\lambda_{\text{Irred}}(\psi) = 0$. If ψ is a completely pruned tangle of order 0, then ψ is a closed walk and has $\lambda_{\text{Irred}}(\psi) = 1$.

If $d = 4$, then consider the tangle of order 1 with three vertices, consisting of one "middle" vertex joined by two edges to each of two vertices. (This is a tangle by labelling the left to middle edges and the middle to right edges π_1, π_2.) We easily compute $\lambda_{\text{Irred}} = \sqrt{3}$, as this graph is bipartite and the number of irreducible walks of length $2m$ from the middle vertex, all such walks being closed walks, is clearly $4 \cdot 3^{m-1}$. So for $d = 4$, $\tau_{\text{fund}} = 1$.

For $d = 6$, consider the tangle, $\psi = (V, E)$ with $V = \{v_1, v_2, v_3\}$ with three edges connecting v_1 to v_2 (labelled π_1, π_2, π_3) and two edges connecting v_2 to v_3 (labelled π_1, π_2). We claim that $\lambda_{\text{Irred}}(\psi) > \sqrt{5}$. Say that a closed walk about v_2 ends in "state A" if the last vertex before v_2 was v_1, and otherwise in "state B" (i.e. the second to last vertex is v_3). From state A, taking two additional irreducible steps, there are 4 ways to reach another state A, and two ways to reach another state B. From state B, taking two irreducible steps, there is one way to reach another state B and six ways to reach another state A. It easy follows that $\lambda_{\text{Irred}}(\psi)$ is the square root of the largest eigenvalue of

$$(30) \qquad \begin{bmatrix} 4 & 2 \\ 6 & 1 \end{bmatrix}.$$

But λ_1 of this matrix is $(5 + \sqrt{57})/2$, and this λ_1 is just $\lambda_{\text{Irred}}(\psi)$. It follows that $\lambda_{\text{Irred}}(\psi) > \sqrt{6} > \sqrt{5}$, and hence $\tau_{\text{fund}} \le 2$.

We wish to rule out $\tau_{\text{fund}} = 1$ when $d = 6$. Since we are considering only completely pruned graphs, ψ, each vertex has degree ≥ 2. Such a graph, ψ, of order 1 has all vertices of degree 2 except for one of degree 4 or two of degree 3. In the case where there are vertices, u, v, of degree 3, therefore joined by three disjoint beaded paths, then $\lambda_{\text{Irred}}(\psi)$ is greatest when the beaded paths are each of length 1 (by setting up an obvious map from irreducible closed walks about u from the general graph to the one with beaded paths of length 1); hence $\lambda_{\text{Irred}}(\psi) \le 2$ in this case, since the graph of two vertices joined by three edges has $\lambda_{\text{Irred}} = 2$. Similarly, in the case with u of degree 4, therefore having two beaded closed walks from u, $\lambda_{\text{Irred}}(\psi)$ is greatest when the lengths of the two closed walks are two (they cannot be one since $\mathcal{H}_{n,d}$ does not permit self-loops); hence $\lambda_{\text{Irred}}(\psi) \le \sqrt{3}$ in this case. Hence $\tau_{\text{fund}} > 1$ and therefore $\tau_{\text{fund}} = 2$.

We conclude that $\tau_{\text{fund}} = \lceil \sqrt{d-1} \rceil - 1$ also when $d = 4, 6$. $\square$

4. DETERMINING τ_{fund} FOR $\mathcal{H}_{n,d}$, $\mathcal{I}_{n,d}$, AND $\mathcal{J}_{n,d}$

THEOREM 6.11. *For the model $\mathcal{I}_{n,d}$, we have $\tau_{\text{fund}} = \lceil \sqrt{d-1}\, \rceil - 1$ for all $d \geq 3$.*

PROOF. We argue as with $\mathcal{H}_{n,d}$. The only difference is that in $\mathcal{I}_{n,d}$, two vertices can have as many as d edges between them in a tangle (as opposed to $d/2$ edges in an $\mathcal{H}_{n,d}$ tangle). So the argument in the previous theorem shows that the two-vertex tangles give that $\tau = \lceil \sqrt{d-1}\, \rceil - 1$ equals τ_{fund} provided that $\tau + 2 \leq d$ (as opposed to $\tau + 2 \leq d/2$ for $\mathcal{H}_{n,d}$). But we easily verify that

$$\lceil \sqrt{d-1}\, \rceil + 1 \leq d$$

for all $d \geq 3$ (indeed, we have equality for $d = 3$, and each time we increase $d \geq 3$ by one, $\sqrt{d-1}$ increases by less than one). □

THEOREM 6.12. *For the model $\mathcal{J}_{n,d}$, we have $\tau_{\text{fund}} = \lceil \sqrt{d-1}\, \rceil - 1$ for all $d \geq 3$.*

PROOF. As in $\mathcal{I}_{n,d}$, for any $\tau \leq d - 2$ there is a tangle G_τ that is a pair of vertices with $\tau + 2$ edges joining them. G_τ has order τ and $\lambda_{\text{Irred}} = \tau + 1$; since when $\tau + 2 = d$ we have $\lambda_{\text{Irred}}(G_\tau) = d - 1 \geq \sqrt{d-1}$ giving a supercritical tangle, we need worry only about whether or not there is a tangle of order $\tau \leq d - 2$ that can beat the λ_{Irred} of G_τ. Again, as with $\mathcal{I}_{n,d}$, Lemma 6.9 can be applied to graphs with half-loops, and so by the same argument as for $\mathcal{I}_{n,d}$ we have that only graphs on one or two vertices can possibly beat G_τ. So consider a graph on vertices u, v with a half-loops about u, c half-loops about v, and b edges from u to v. An irreducible path traverses edges of four different states: (1) half-loops about u, (2) edges from u to v, (3) edges from v to u, and (4) half-loops about v. Now we write a transition matrix about the states: for example in state (1) we may either continue on one of $a - 1$ half-loops in state (1) or continue on one of b edges in state (2). We find the transition matrix

$$\begin{bmatrix} a-1 & b & 0 & 0 \\ 0 & 0 & b-1 & c \\ a & b-1 & 0 & 0 \\ 0 & 0 & b & c-1 \end{bmatrix},$$

and λ_{Irred} of our graph is this matrice's largest eigenvalue. The order of the graph is $a + b + c - 2$ (recall, each half-loop contributes one to the order of a graph). But the row sum is never greater than $a + b + c - 1$ (and always less unless a or c vanishes), and so if this graph has order τ its λ_{Irred} is no more than $\tau + 1$. Hence no $\mathcal{J}_{n,d}$ tangle of order τ beats G_τ, provided that $\tau \leq d - 2$. Thus τ_{fund} is the smallest number with $\tau_{\text{fund}} + 1 \geq \sqrt{d-1}$. □

CHAPTER 7

Ramanujan Functions

In this section we discuss Ramanujan functions in order to (1) explain their significance, and (2) give some intuition on some very technical issues surrounding the asymptotic expansion for irreducible traces (as in Chapter 8).

DEFINITION 7.1. *A function, $f(k)$, on positive integers, k, is said to be d-Ramanujan of order $\alpha > 0$ if there is a polynomial $p = p(k)$ and a constant $c > 0$ such that*
$$|f(k) - (d-1)^k p(k)| \le c k^c \alpha^k$$
for all k. We call $(d-1)^k p(k)$ the principal term *of f, and $f(k) - (d-1)^k p(k)$ the* error term *(both terms are uniquely determined if $\alpha < d - 1$). A function is* super-d-Ramanujan *if it is d-Ramanujan of order 1.*

A d-Ramanujan function as defined before, in Definition 2.1, is just a d-Ramanujan of order $\sqrt{d-1}$.

Let $N(k)$ be the number of irreducible cycles of length k in a d-regular graph. Then in [**LPS86**] it is shown that if $N(k)$ is d-Ramanujan, then any eigenvalue, $\lambda \ne \pm d$, of the graph satisfies $|\lambda| \le 2\sqrt{d-1}$. The discussion there also shows that in any case, if λ is the eigenvalue of largest absolute value $< d$, then $N(k)$ is d-Ramanujan of order α with
$$\alpha = \frac{|\lambda| + \sqrt{\lambda^2 - 4(d-1)}}{2}$$
(and not for any smaller an α). Any discussion of irreducible traces and eigenvalues is bound to be tied to d-Ramanujan functions.

One important property of d-Ramanujan functions of order α is that they are closed under addition. Another very important property is that they are closed under *convolution*, which we now formally explain. This property will be used in Chapter 14, and refined versions of it will be used in Chapter 8.

THEOREM 7.2. *Let f_1, f_2 be d-Ramanujan of order α with $\alpha < d - 1$. Then their convolution,*
$$g(k) = (f_1 * f_2)(k) = \sum_{j=1}^{k-1} f_1(j) f_2(k-j)$$
is also d-Ramanujan of order α.

The techniques in Chapter 8 prove a more precise version of this theorem (keeping track of the sizes of the the error term and the coefficients of the principal part); for this reason we keep the argument below concise.

PROOF. For $i = 1, 2$ let
$$f_i(k) = (d-1)^k p_i(k) + e_i(k)$$
where p_i are polynomials and the $|e_i(k)|$ are bounded by $ck^c \alpha^k$ for some k. We may also write
$$f_i(k) = (d-1)^k \big(p_i(k) + \tilde{e}_i(k)\big), \qquad \text{where } \tilde{e}_i(k) = (d-1)^{-k} e_i(k).$$
Since convolution is bilinear, we easily see
$$f * g = e_1 * e_2 + (d-1)^k (p_1 * p_2 + p_1 * \tilde{e}_2 + p_2 * \tilde{e}_1).$$
It suffices to show that
$$e_1 * e_2, \quad (d-1)^k (p_1 * p_2)(k), \quad (d-1)^k (p_1 * \tilde{e}_2)(k), \quad (d-1)^k (p_2 * \tilde{e}_1)(k)$$
are d-Ramanujan of order α.

According to Sublemma 2.15 of [**Fri91**], $p_1 * p_2$ is a polynomial. Next
$$(p_1 * \tilde{e}_2)(k) = \sum_{j=1}^{k-1} (d-1)^{-j} p_1(k-j) e_2(j) = \Sigma_1 - \Sigma_2$$
where
$$\Sigma_1 = \sum_{j=1}^{\infty} p_1(k-j)(d-1)^{-j} e_2(j),$$
$$\Sigma_2 = \sum_{j=k}^{\infty} p_1(k-j)(d-1)^{-j} e_2(j).$$
Writing
$$p_1(k-j) = \sum a_{r,s} k^r j^s,$$
we see that Σ_1 is a polynomial, and Σ_2 is bounded by $ck^c \alpha^k (d-1)^{-k}$ (see Chapter 8, especially Lemma 8.8, for details). This shows $(d-1)^k (p_1 * \tilde{e}_2)$ is d-Ramanujan of order α. Similarly, so is $(d-1)^k (p_2 * \tilde{e}_1)$; $e_1 * e_2$ is easily also seen to be so (with zero principal term). □

CHAPTER 8

An Expansion for Some Selective Traces

In this section we prove the first crucial expansion theorem. Our second such theorem, Theorem 9.3, will extend these ideas.

THEOREM 8.1. *Let r be a positive integer, and let Ψ be a set of tangles containing all supercritical tangles of order less than r. Then there is an $S_0 = S_0(r)$ such that for all $S \geq S_0$ the following holds. We have*

$$E[\mathrm{IrSelTr}_{S,\Psi}(G;k)] = f_0(k) + \frac{f_1(k)}{n} + \cdots + \frac{f_{r-1}(k)}{n^{r-1}} + \frac{\mathrm{error}}{n^r},$$

where the f_i are d-Ramanujan and the error term satisfies the bound given in Theorem 5.9.

We begin by explaining why this theorem is an easy consequence of Theorem 5.9 and the following theorem.

THEOREM 8.2. *Fix a lettering, $\mathcal{L}$, of type T, and fixed non-negative integers $\ell_1, \ldots, \ell_{d/2}$. Let*

$$R_{T,\mathcal{L}}(k_1, \ldots, k_b) = \sum_{(w_1, \ldots, w_b)} \prod_{j=1}^{d/2} \bigl(a_j(w_1) + \ldots + a_j(w_b)\bigr)^{\ell_j}$$

where the sum is over all tuples of words $(w_1, \ldots, w_b)$ such that each w_i is irreducible and of length k_i and is compatible with $\mathcal{L}$. Let $\widetilde{T}$ be a B-new type based on T, and let

$$(31) \qquad f(k) = \sum_{m_i \geq 1} \sum_{\substack{k_1 m_1 + \ldots + k_b m_b = k \\ k_i \geq B \text{ if } e_i \in E_{\mathrm{long}} \\ k_i = k_i^{\mathrm{fixed}} \text{ if } e_i \in E_{\mathrm{fixed}}}} W_{\widetilde{T}}(\vec{m}; S, \Psi) R_{T,\mathcal{L}}(k_1, \ldots, k_b)$$

(with W as in Theorem 6.6). Then f is d-Ramanujan for all $B \geq B_0 = B_0(T)$.

Assume Theorem 8.2 for the moment. Let $\mathcal{W}$ (respectively, $\mathcal{W}_T$ and $\mathcal{W}_{\widetilde{T}}$) be the walk collections corresponding to irreducible, (S, Ψ)-selective walks (respectively, and that are associated to the type T and $\widetilde{T}$). These walk collections are all SSIIC. According to Theorem 5.9

$$\mathrm{WalkSum}(\mathcal{W}, k, n) = f_0(k) + \frac{f_1(k)}{n} + \cdots + \frac{f_{r-1}(k)}{n^{r-1}} + \frac{\mathrm{error}}{n^r},$$

where

$$(32) \qquad f_i(k) = \sum_{j=0}^{r-1} \sum_{(w;[\vec{t}]) \text{ order } j, \in \mathcal{W}(k)} p_{i-j}\bigl(a_1(w;[\vec{t}]), \ldots, a_d(w;[\vec{t}])\bigr).$$

It suffices to show that these f_i are d-Ramanujan. We know there are finitely many types of order at most $r-1$. Now fix S_0 to be the max over $B_0(T)$ over types, T, of order at most $r-1$ (and B_0 as in Theorem 8.2). For any $S \geq S_0$, choose $B = S+1$; of course, $B \geq B_0(T)$ for any type, T, of order at most $r-1$. We know there are finitely many B-new types based on any type. So the sum involving $\mathcal{W}$ in equation (32) decomposes as a finite sum over $\mathcal{W}_T$'s or $\mathcal{W}_{\widetilde{T}}$'s. Furthermore, each expansion polynomial p_{i-j}, involving a fixed new type, $\widetilde{T}$, is just a function of $a_1, \ldots, a_{d/2}$ over appropriate forms, and each a_i of the form is just the sum of the a_i along each edge of the form. Therefore Theorem 8.2 just says that each f_i, when summing over a $\mathcal{W}_{\widetilde{T}}$, is d-Ramanujan. Summing over all $\mathcal{W}_{\widetilde{T}}$ shows that the f_i corresponding to $\mathcal{W}$ are also d-Ramanujan.

Proof (of Theorem 8.2) Clearly it suffices to prove the following theorem with $R_{T,\mathcal{L}}$ replaced by

$$\sum_{(w_1,\ldots,w_b)} \prod_{i=1}^{b} \prod_{j=1}^{d/2} a_j^{\ell_{ij}}(w_i),$$

with ℓ_{ij} any set of non-negative integers.

Our Lemma 5.10 reduces the above theorem to the following.

THEOREM 8.3. *With notation as in Theorem 8.2, let K_1, K_2, K_3 be a partition of $k_1, \ldots, k_b$, and let $|K_i|$ for $i = 1, 2, 3$ denote the sum of the k_j in K_i. Then for fixed non-negative integers $\ell_1, \ldots, \ell_b$, Theorem 8.2 holds with $R_{T,\mathcal{L}}$ replaced by*

(33) $$R_{T,\mathcal{L}}(k_1,\ldots,k_b) = (d-1)^{|K_1|}(-1)^{|K_2|}k_1^{\ell_1} \cdots k_b^{\ell_b}.$$

More generally, Theorem 8.2 holds with $R_{T,\mathcal{L}}$ replaced by

(34) $$R_{T,\mathcal{L}}(k_1,\ldots,k_b) = (d-1)^{|K_1|}k_1^{\ell_1} \cdots k_u^{\ell_u} \beta(k_{u+1},\ldots,k_b),$$

where the edges are ordered so that

$$\{i | e_i \in E_{\text{long}} \text{ and } k_i \in K_1\} = \{1,\ldots,u\},$$

and where β is a function such that

$$|\beta(k_{u+1},\ldots,k_b)| \leq c(|k_{u+1}| + \cdots + |k_b|)^c$$

for some constant c.

The R of equation (33) is all that is needed for $\mathcal{G}_{n,d}$; it will be convenient (if not necessary) to use the R of equation (34) for $\mathcal{J}_{n,d}$ (see Chapter 14).

PROOF. It suffices to deal with the R of equation (34). Let

$$\widetilde{K}_i = \{k_j \in K_i \mid e_j \in E_{\text{long}}\}.$$

We may assume $E_{\text{long}} = \{e_1,\ldots,e_t\}$ and $\widetilde{K}_1 = \{k_1,\ldots,k_u\}$; set

$$M_1 = m_1 + \cdots + m_t, \qquad M_2 = m_{t+1} + \cdots + m_b,$$
$$j_1 = k_1 m_1 + \cdots + k_t m_t, \qquad \text{and } j_2 = k_{t+1}m_{t+1} + \cdots + k_b m_b.$$

Clearly it suffices to prove the theorem for

$$f(k) = \sum_{\vec{m}} W_{\widetilde{T}}(\vec{m}; S, \Psi) \sum_{\substack{k_1 m_1 + \cdots + k_b m_b = k \\ k_i \geq B \text{ for } i \leq t}} (d-1)^{|\widetilde{K}_1|} k_1^{\ell_1} \ldots k_u^{\ell_u} \beta(k_{u+1},\ldots,k_t),$$

understanding that $k_{t+1}, \ldots, k_t$ are fixed by $\vec{k}^{\text{fixed}}$.

8. AN EXPANSION FOR SOME SELECTIVE TRACES

DEFINITION 8.4. *The (coefficient) norm, $|p|$, of a polynomial, p (which is possibly multivariate), is the largest absolute value among its coefficients.*

Working with this notion of a norm is a bit "weak," (i.e., sometimes much stronger statements would hold with other norms), but this notion is sufficient for our purposes.

Let
$$f_{\vec{m}}(k) = \sum_{\vec{k} \cdot \vec{m} = k} (d-1)^{|\widetilde{K}_1|} k_1^{\ell_1} \ldots k_u^{\ell_u} \beta(k_{u+1}, \ldots, k_t).$$

THEOREM 8.5. *For any vector of positive integers, $\vec{m}$, we have $f_{\vec{m}}$ is d-Ramanujan with principal term $(d-1)^k p_{\vec{m}}(k)$ and error term $e_{\vec{m}}(k)$ satisfying*
$$|p_{\vec{m}}| \leq c(d-1)^{(-BM_1 - M_2 + c)/2},$$

and
$$|e_{\vec{m}}(k)| \leq ck^c (d-1)^{(k - BM_1 - M_2 + Bc)/2},$$

where c depends only on the ℓ_i and β.

PROOF. Fix a value of $\vec{m}$. Without loss of generality we may assume $m_1 = \cdots = m_s = 1$ and $m_{s+1}, \ldots, m_u \geq 2$. For now assume that $s \geq 1$; we will later indicate the minor changes needed for the situation $s = 0$ (i.e. when there are no m_i belonging to $\widetilde{K}_1$ that equal one). Let
$$g_{\vec{m}}(r) = \sum_{\substack{k_{s+1}, \ldots, k_b \text{ s.t.} \\ k_{s+1}m_{s+1} + \cdots + k_b m_b = r \\ k_i \geq B \text{ for } i \leq t}} (d-1)^{k_{s+1} + \cdots + k_u} \beta(k_{u+1}, \ldots, k_t).$$

LEMMA 8.6. *If $\widetilde{T}$ is a B-new type, then*
$$|g_{\vec{m}}(r)| \leq cr^c (d-1)^{(r - BM_1 - M_2 + Bc)/2}$$

for some constant c depending only on $\widetilde{T}$.

PROOF. Since each k_i is at most r,
$$\beta(k_{u+1}, \ldots, k_t)$$
is bounded by cr^c, and it suffices to prove the estimate for $g_{\vec{m}}$ replaced with
$$\sum_{\substack{k_{s+1}m_{s+1} + \cdots + k_b m_b = r \\ k_i \geq B \text{ for } i \leq t}} (d-1)^{k_{s+1} + \cdots + k_u}.$$

But there are only $\binom{r+b-s-1}{b-s-1}$ ways of writing r as the sum of $b-s$ positive integers. So it suffices to show
$$(d-1)^{k_{s+1} + \cdots + k_u} \leq (d-1)^{(r - BM_1 - M_2 - Bc)/2}.$$

Now we have
$$r = k_{s+1}m_{s+1} + \cdots + k_b m_b,$$
so
$$2k_{s+1} + \cdots + 2k_u = r - (k_{s+1}m_{s+1} + \cdots + k_b m_b) + (2k_{s+1} + \cdots + 2k_u)$$
$$= r - (m_{s+1} - 2)k_{s+1} - \cdots - (m_u - 2)k_u - m_{u+1}k_{u+1} - \cdots - m_b k_b.$$

Since $m_i \geq 2$ for i between $s+1$ and u, and since $k_i \geq B$ for $i \leq t$ (and $k_i \geq 1$ for all i), we conclude

$$2k_{s+1} + \cdots + 2k_u \leq r - (m_{s+1} - 2)B - \cdots - (m_u - 2)B$$
$$-m_{u+1}B - \cdots - m_t B - m_{t+1} - \cdots - m_b$$
$$= r - (m_{s+1} + \cdots + m_t)B + 2(u-s)B - (m_{t+1} + \cdots + m_b)$$
$$= r - \big(M_1 - s - 2(u-s)\big)B - M_2.$$

Hence

$$k_{s+1} + \cdots + k_u \leq (2k_{s+1} + \cdots + 2k_u)/2 \leq \Big(r - \big(M_1 - s - 2(u-s)\big)B - M_2\Big)/2.$$

But u, s are bounded by the number of edges in $\widetilde{T}$. □

We will need another lemma.

LEMMA 8.7. *For any non-negative integers $\ell_1, \ldots, \ell_s$ there is a polynomial Q such that for all $k \geq s$ we have*

$$\sum_{\substack{k_1 + \cdots + k_s = k \\ \text{integers } k_i \geq 1}} k_1^{\ell_1} \ldots k_s^{\ell_s} = Q(k).$$

PROOF. This is a special case of Sublemma 2.15 of [**Fri91**] (proven in a straightforward induction on s). □

Now let

$$j_{11} = k_1 m_1 + \cdots + k_s m_s = k_1 + \cdots + k_s, \qquad j_{12} = k_{s+1} m_{s+1} + \cdots + k_t m_t$$

(so that $j_{11} + j_{12} = j_1$). In the notation of the above two lemmas, letting $j' = j_{12} + j_2$, we have

$$f_{\vec{m}}(k) = \sum_{j_{11} + j' = k} (d-1)^{j_{11}} Q(j_{11}) g_{\vec{m}}(j')$$

(35)
$$= \sum_{r=1}^{k-s} (d-1)^{k-r} Q(k-r) g_{\vec{m}}(r);$$

here we sum until $r = k - s$ since Lemma 8.7 requires $k \geq s$ (and $Q(k-r)$ vanishes for $k < s$), and we sum from $r = 1$ to simplify the expression, despite the fact that $g_{\vec{m}}(r)$ clearly vanishes for

$$r < B(m_{s+1} + \cdots + m_t) + m_{t+1}k_{t+1} + \cdots + m_b k_b.$$

The sum in equation (35) is clearly $\Sigma_1(k) - \Sigma_2(k)$, where

$$\Sigma_1(x) = \sum_{r=1}^{\infty} (d-1)^{k-r} Q(x-r) g_{\vec{m}}(r),$$
$$\Sigma_2(x) = \sum_{r=k-s+1}^{\infty} (d-1)^{k-r} Q(x-r) g_{\vec{m}}(r),$$

assuming these sums converge.

We claim $\Sigma_1(k)$ will be the principal part of $f_{\vec{m}}(k)$, and $\Sigma_2(k)$ will be the error term. First we need the following lemma.

LEMMA 8.8. *For any positive integer, D, there is a C_2 such that the following holds. Let $g(r)$ be a function defined on non-negative integers, r, such that $|g(r)| \leq C_1 r^D \rho^r$, with $\rho < 1$. Let $Q = Q(x)$ be any polynomial of degree at most D. Then (1) the infinite sum*

$$h(x) = \sum_{r=1}^{\infty} Q(x-r)g(r)$$

is convergent (in coefficient norm), (2) the degree of h is that of Q, and (3) we have

$$|h| \leq C_1 C_2 (1-\rho)^{-2D} |Q|.$$

The same is true for the sum

$$h_u(x) = \sum_{r=u+1}^{\infty} Q(x-r)g(r),$$

for any positive integer u, except that we replace the last claim with the estimate

$$|h_u| \leq C_1 C_2 u^{2D} (1-\rho)^{-2D} \rho^u |Q|$$

(however, the C_2 in this equation might need to be larger than that in the estimate for h).

PROOF. First we observe that since

$$\sum_{r=0}^{\infty} \binom{r}{j} \rho^r = \frac{\rho^{j+1}}{(1-\rho)^j},$$

we have

$$\sum_{r=0}^{\infty} r^j \rho^r = \frac{q_j(\rho)}{(1-\rho)^j},$$

where q_j is some polynomial of degree at most $j+1$.

We first prove the claims on h. By the binomial theorem, and the fact that Q's degree is bounded, it suffices to examine only the cases where $Q(x-r)$ is replaced by $x^i r^j$ for $i + j \leq D$. In this case $h(x)$ becomes

$$\sum_{r=1}^{\infty} x^i r^j g(r) = x^i \sum_{r=1}^{\infty} r^j g(r),$$

and we have

$$\sum_{r=1}^{\infty} r^j |g(r)| \leq \sum_{r=0}^{\infty} r^j C_1 r^D \rho^r = \frac{C_1 \rho^{j+D+1}}{(1-\rho)^{j+D}}.$$

This establishes the claim on h. The claim on h_u is reduced to h via

$$h_u(x) = \sum_{r=1}^{\infty} \tilde{Q}(x-r) \tilde{g}(r),$$

where $\tilde{Q}(x) = Q(x-u)$ and $\tilde{g}(r) = g(r+u)$. So $\tilde{g}$ satisfies the same estimate as does g, except with an extra factor of $(r+u)^D r^{-D} \rho^u \leq C u^D \rho^u$; the binomial theorem implies that $|\tilde{Q}|$ is at most $|Q|u^D$ times a constant depending on D. □

We continue with the proof of Theorem 8.5. We have

$$(d-1)^{-k}\Sigma_1(k) = \sum_{r=1}^{\infty} Q(k-r)[(d-1)^{-r}g_{\vec{m}}(r)] = \sum_{r=1}^{\infty} Q(k-r)\tilde{g}(r),$$

where $\tilde{g}(r) = (d-1)^{-r}g_{\vec{m}}(r)$. Now Q is fixed in the theorem, so $|Q|$ can be regarded as a constant. Also, since

$$|g_{\vec{m}}(r)| \le cr^c(d-1)^{(r-BM_1-M_2+Bc)/2},$$

according to Lemma 8.6, we have

$$|\tilde{g}(r)| \le cr^c(d-1)^{(-BM_1-M_2+Bc)/2}.$$

It follows that $(d-1)^{-k}\Sigma_1(k) = h(k)$, where h is a polynomial with

$$|h| \le c(d-1)^{(-BM_1-M_2+Bc)/2},$$

assuming that $d > 2$ (so that $1-\rho$ with $\rho = (d-1)^{-1/2}$ is strictly positive).

Furthermore, Lemma 8.8 also implies that

$$|\Sigma_2| \le c(d-1)^k(k-s+1)^{2D}(d-1)^{-(k-s)/2}(d-1)^{(-BM_1-M_2+Bc)/2}$$
$$\le c'k^{2D}(d-1)^{(-BM_1-M_2+Bc+k)/2}.$$

Now we see that $\Sigma_1(k) + \Sigma_2(k)$ is the decomposition of $f_{\vec{m}}(k)$ into principal and error terms, as claimed before, and that these terms satisfy the bounds stated in Theorem 8.5.

Finally we indicate the minor changes when $s = 0$. In this case we take Q to be the function $Q(0) = 1$ and Q vanishing elsewhere. Then $f_{\vec{m}}(k) = g_{\vec{m}}(k)$, so Lemma 8.6 shows that $f_{\vec{m}}$ is d-Ramanujan with zero principal part. □

We continue with the proof of Theorem 8.3. We are studying

$$f(k) = \sum_{\vec{m}} W_{\widetilde{T}}(\vec{m}; S, \Psi) f_{\vec{m}}(k).$$

Set

$$F(M_1, M_2; k) = \sum_{\substack{m_1+\ldots+m_t=M_1 \\ m_{t+1}+\ldots+m_b=M_2}} W_{\widetilde{T}}(\vec{m}; S, \Psi) f_{\vec{m}}(k),$$

so that

$$f(k) = \sum_{M_1, M_2 > 0} F(M_1, M_2; k).$$

Theorem 6.6 combined with Theorem 8.5 gives that for fixed M_1, M_2 we have that $F(M_1, M_2; k)$ is d-Ramanujan with principle term $(d-1)^k P_{M_1,M_2}(k)$ and error term $E_{M_1,M_2}(k)$ where

$$|P_{M_1,M_2}| \le (d-1)^{(-BM_1-M_2+c)/2}c(\sqrt{d-1}-\epsilon)^{BM_1+M_2},$$

$$|E_{M_1,M_2}(k)| \le ck^c(d-1)^{(k-BM_1-M_2+Bc)/2}c(\sqrt{d-1}-\epsilon)^{BM_1+M_2},$$

and the degree of P_{M_1,M_2} is bounded independent of M_1, M_2. So we sum over all M_1, M_2 to conclude that f is d-Ramanujan. □

CHAPTER 9

Selective Traces In Graphs With (Without) Tangles

Let us review our general approach to the Alon conjecture. We are interested in expansions in $1/n$ of the expected value of the k-th irreducible trace. Unfortunately these expansions have some coefficients that fail to be d-Ramanujan, and, as explained in Chapter 2, this prevents us from proving the Alon conjecture. Replacing irreducible traces with selective traces gives d-Ramanujan coefficients up to any desired power of $1/n$. However, selectivity, in the presence of appropriate tangles, modifies the irreducible trace in a way that seems hard to control. Thus we don't know how to use the results of the last section to conclude the Alon conjecture.

The last main idea of the proof is to get an expansion for the expected value of selective traces counted only when appropriate tangles are present (i.e., the selective trace multiplied by a characteristic function over those graphs in $\mathcal{G}_{n,d}$ with appropriate tangles). The methods of the last section generalize, rather tediously, to such expansions. These expansions will also have d-Ramanujan coefficients up to any desired power of $1/n$. It follows that we also get such expansions for the expected value of the selective trace counted only when appropriate tangles are *not* present; for this count, the selective trace and irreducible trace are the same. This information turns out to be enough to prove the Alon conjecture (with an auxiliary lemma proven in Chapter 11).

Before doing the above, it is crucial to know that a certain set of tangles is finite.

DEFINITION 9.1. *Recall that Ψ_{eig} is the set of supercritical tangles, i.e., whose λ_{Irred} is at least $\sqrt{d-1}$. Recall that $\Psi_{\mathrm{eig}}[r]$ is the subset of elements of Ψ_{eig} of order at most r. Let $\Psi_{\mathrm{min}}[r]$ be the set of tangles of $\Psi_{\mathrm{eig}}[r]$ that are minimal with respect to inclusions, i.e., that don't have another element of $\Psi_{\mathrm{eig}}[r]$ properly included in it.*

LEMMA 9.2. *The set $\Psi_{\mathrm{min}}[r]$ is finite.*

This means that containing a supercritical tangle of order at most r is equivalent to containing one of a finite set of tangles.

PROOF. Assume that $\Psi_{\mathrm{min}}[r]$ is not finite. With each tangle we associate a type which is the labelled graph obtained by suppressing the degree two vertices. Since there are finitely many types of order at most r, there must be an infinite number of $\Psi_{\mathrm{min}}[r]$ tangles of some type, T. By passing to a subsequence we may assume there is an infinite sequence of $\Psi_{\mathrm{min}}[r]$ elements, $\{\psi_i\}$, such that for each edge of T the associated labelling is either constant or has length tending to infinity; furthermore, the length must tend to infinity along at least one T edge. Let ψ_∞ be the limiting tangle, where we discard all edges with length tending to infinity.

We claim $\lambda_{\text{Irred}}(\psi_i) = \lambda_1(T^i_{\text{Irred}})$, where T^i_{Irred} is the VLG with underlying directed graph T_{Irred}, and where $e = (v_1, v_2) \in E_{T_{\text{Irred}}}$ has length equal to the length, $\ell(v_1)$, of v_1 in ψ_i (recall v_1 can be viewed as a directed edge of T); indeed, with a vertex path $v_1, \ldots, v_r$ in T_{Irred} with $v_j = (u_j, u_{j+1})$ for $u_j \in V_T$, we associate the walk $u_1, \ldots, u_{r+1}$, which has ψ_i length equal to the sum of the $\ell(v_i)$. For a closed walk, where $u_{r+1} = u_1$, its length (in T^i_{Irred}) $\ell(v_1) + \cdots + \ell(v_r)$, which corresponds to a unique $(\psi_i)_{\text{Irred}}$ closed walk of the same length, arising from the subdivided v_j. This correspondence is clearly a length preserving bijection between T^i_{Irred} closed walks and $(\psi_i)_{\text{Irred}}$ closed walks. Hence $\lambda_{\text{Irred}}(\psi_i) = \lambda_1(T^i_{\text{Irred}})$.

Similarly $\lambda_{\text{Irred}}(\psi_\infty) = \lambda_1(T^\infty_{\text{Irred}})$ with T^∞_{Irred} defined similarly. Now by Theorem 3.6, $\lambda_1(T^i_{\text{Irred}}) \to \lambda_1(T^\infty_{\text{Irred}})$, and so $\lambda_{\text{Irred}}(\psi_\infty) \geq \sqrt{d-1}$. Also ψ_∞'s order is less than that of the ψ_i (because of the edge removal(s)). Hence ψ_∞ is again a $\Psi_{\text{eig}}[r]$ tangle. But ψ_∞ properly contains (all) ψ_i, which contradicts the supposed minimality of the ψ_i. Hence $\Psi_{\min}[r]$ is finite. □

We illustrate the above lemma with an example. Let ψ_i be a sequence of tangles whose underlying graph is the same except for one beaded cycle of length i about some vertex. This infinite collection of tangles would prove troublesome to the methods of this section. However either (1) $\lambda_{\text{Irred}}(\psi_i) < \sqrt{d-1}$ for some i, at which point only finitely many of the ψ_i are relevant, or (2) the limiting tangle, ψ_∞, has $\lambda_{\text{Irred}}(\psi_\infty) \geq \sqrt{d-1}$, in which case a ψ_i inclusion implies a ψ_∞ inclusion.

THEOREM 9.3. *Let Ψ be a finite set of pruned (nonempty) tangles of order at least 1. Let χ_Ψ be the indicator function of the event that $G \in \mathcal{G}_{n,d}$ contains a (i.e., at least one) tangle from Ψ, i.e.,*

$$\chi_\Psi(G) = \begin{cases} 1 & \text{if } G \text{ contains a tangle from } \Psi, \\ 0 & \text{if not.} \end{cases}$$

Let Ψ' be a set of tangles including all supercritical tangles of order less than r. Then for any r there is an $S_0 = S_0(r, \Psi, \Psi')$ such that for all $S \geq S_0$ we have an expansion

$$(36) \quad E[\chi_\Psi \text{IrSelTr}_{S,\Psi'}(G; k)] = f_0(k) + \frac{f_1(k)}{n} + \cdots + \frac{f_{r-1}(k)}{n^{r-1}} + \frac{\text{error}}{n^r},$$

where the f_i are d-Ramanujan and the error term satisfies

$$|\text{error}| \leq c k^{\tilde{r}} (d-1)^k$$

with c and $\tilde{r}$ depending only on r, Ψ, and Ψ'.

We believe this theorem is true even if Ψ contains cycles, i.e., tangles of order 0. But to prove this would be more difficult, since the last part of Lemma 9.6 would not be true (see the remark that follows this lemma).

Before giving the proof of Theorem 9.3, we give an important corollary of it and Theorem 8.1.

COROLLARY 9.4. *With notation and conditions as in Theorem 9.3, we have that*

$$E[(1 - \chi_\Psi) \text{IrSelTr}_{S,\Psi'}(G; k)]

also has an expansion of the form given by the right-hand-side of equation (36).

PROOF. **(of Theorem 9.3)** We want to generalize walk sums, forms, types, etc. into structures that incorporate the presence of a Ψ tangle. We shall first explain why we run into inclusion/exclusion and tangle automophisms (as in Theorem 4.7).

Define a *potential tangle specialization* as a pair, (Ω, σ), of a tangle, Ω, and an inclusion $\sigma \colon V_\Omega \to \{1, \ldots, n\}$. Let $\chi_{\Omega, \sigma}$ denote the indicator function of the event that Ω is contained in $G \in \mathcal{G}_{n,d}$ via the map σ. Let $\chi_{(w;\vec{t})}$ be the indicator function of the event, $\mathcal{E}(w; \vec{t})$, of the potential walk $(w; \vec{t})$. We plan to study sums involving various $\chi_{(w;\vec{t})} \chi_{\Omega,\sigma}$.

For now consider $\chi_{\Omega,\sigma}$ alone. Since the word "form" was used earlier to mean "look at the graph traced out and forget the specific values of the vertices," it makes sense to view Ω itself as the "form" of (Ω, σ), and let

$$\mathrm{E}\,[\Omega]_n = \sum_\sigma \mathrm{E}\,[\chi_{\Omega,\sigma}] = \frac{n!}{(n-v)!} \prod_{i=1}^{d/2} \frac{(n-a_i)!}{n!},$$

where, as usual, the a_i are the number of Ω edges labelled π_i and v is the number of vertices in Ω. The problem is that $\mathrm{E}\,[\Omega]_n$ gives the expected number of times Ω is included into a random graph; when computing traces and walk sums, we do wish to count; but when looking at tangle inclusions, when don't wish to count—we want to compute only the function χ_Ψ, i.e., we want a 1 when a Ψ tangle is present, and a 0 otherwise.

Given two tangles, ψ, ψ', let $N(\psi, \psi')$ denote the number of inclusions of ψ into ψ'. Of course, $N(\psi, \psi)$ is the number of automorphisms of ψ. Also, $\mathrm{E}\,[\Omega]_n$ is just the expected value of $N(\Omega, G)$ for a $G \in \mathcal{G}_{n,d}$ (viewing G as a tangle). Let $\psi \le \psi'$ (respectively, $\psi < \psi'$) denote that ψ has an inclusion (respectively, proper inclusion) into ψ'. By a Ψ-tangle we mean any tangle isomorphic to an element of Ψ. A *derived tangle of* Ψ is a tangle that is the nonempty union of Ψ-tangles. Let Ψ^+ be a collection of one derived tangle of Ψ in every tangle isomorphism class.

PROPOSITION 9.5. *There exist reals numbers* $\{\mu_\Omega\}_{\Omega \in \Psi^+}$ *such that for any* $\Omega' \in \Psi^+$ *we have*

$$\sum_{\Omega \le \Omega'} N(\Omega, \Omega') \mu_\Omega = 1.$$

PROOF. This is just generalized Möbius inversion: Ψ^+ is a partially ordered set, and given Ω' there are only finitely many Ω with $\Omega < \Omega'$. Furthermore $N(\Omega', \Omega')$ is positive for all Ω'. So we can inductively solve for μ_Ω. □

It follows that

$$\text{(37)} \quad \mathrm{E}\,[\chi_\Psi \mathrm{IrSelTr}_{S,\Psi'}(G; k)] = \sum_{(w,\vec{t}) \in \mathcal{W}} \mathrm{E}\left[\chi_{(w;\vec{t})} \chi_\Psi\right]$$

$$\text{(38)} \quad = \sum_{(w,\vec{t}) \in \mathcal{W}} \sum_{(\Omega,\sigma), \Omega \in \Psi^+} \mathrm{E}\left[\chi_{(w;\vec{t})} \chi_{\Omega,\sigma}\right] \mu_\Omega,$$

where $\mathcal{W}$ is the walk collection corresponding to the irreducible (S, Ψ') selective walks.

LEMMA 9.6. *Let $\mathcal{W}_{<r}$ be those elements of $\mathcal{W}$ of order less than r, and similarly for $\Psi^+_{<r}$. In equation (38), by replacing the summation over $\mathcal{W}$ and Ψ^+ by summation over $\mathcal{W}_{<r}$ and $\Psi^+_{<r}$, the difference is at most $C(d-1)^k n^{-r} k^{2r}$. Furthermore, the set $\Psi^+_{<r}$ is finite for each r.*

We remark that if Ψ could contain a tangle of order 0, i.e., one whose underlying graph is a cycle, then $\Psi^+_{<1}$ would be infinite, containing arbitrarily large disjoint unions of tangles whose underlying graph is a cycle.

PROOF. We know that we can ignore $(w; \vec{t}\,)$'s of order at least r by using Theorem 5.9 and the fact that $|\chi_\Psi| \le 1$ in equation (37). This leaves us with a sum over $\mathcal{W}_{<r}$ and Ψ^+, and it is left to see what happens when Ψ^+ is replaced by $\Psi^+_{<r}$.

For the (Ω, σ)'s, notice that every tangle in Ψ^+ of order at least r must contain a tangle of order between r and $r+s-1$, where s is the maximum number of edges in a Ψ-tangle; this follows because taking the union of any graph with a Ψ-tangle increases the number of edges by at most s. Lemma 4.10 shows that there are only finitely many elements of Ψ^+ of order at most $r+s-1$ (and that any such element is the union of at most $r+s-1$ elements of Ψ). Theorem 4.7 now shows that there is a $C = C(r)$ such that a graph contains a Ψ^+-tangle of order at least r with probability at most Cn^{-r}. Let $\Psi^+_{<r}$ be the subset of tangles of order less than r in Ψ^+. Then

$$\chi_\Psi(G) = h(G) + \sum_{\Omega \in \Psi^+_{<r}} N(\Omega, G) \mu_\Omega,$$

where $h(G)$ is a function bounded by a 1 plus the finite sum of all $|\mu_\Omega|$ over $\Omega \in \Psi^+_{<r}$. Truncating the sum in equation (38) to $\Omega \in \Psi^+_{<r}$ therefore introduces an error of at most

$$\sum_{(w,\vec{t}\,) \in \mathcal{W}} \mathrm{E}\left[\chi_{(w;\vec{t}\,)} h(G)\right] \le d(d-1)^{k-1} C n^{-r} \max(h).$$

$\square$

According to Lemma 9.6, it suffices to fix an $\Omega \in \Psi^+_{<r}$ and to show that

(39) $$\sum_{(w,\vec{t}\,) \in \mathcal{W}_{<r}} \sum_{\sigma} \mathrm{E}\left[\chi_{(w;\vec{t}\,)} \chi_{\Omega,\sigma}\right]$$

has an expansion like the right-hand-side of equation (36). So fix an $\Omega \in \Psi^+_{<r}$; we now define the "form" of $(w; \vec{t}\,)$, but we incorporate into the form the information of how Ω, σ overlaps with $(w; \vec{t}\,)$ by allowing Ω and Γ to share vertices and edges.

DEFINITION 9.7. *An Ω-specialization of a form, Γ, is an inclusion $\iota: V_\Gamma \cup V_\Omega \to \{1, \ldots, n\}$. An Ω-isomorphism of forms is an isomorphism of forms $\Gamma_1 \to \Gamma_2$ which is the identity from $V_{\Gamma_1} \cap V_\Omega$ to $V_{\Gamma_2} \cap V_\Omega$. We introduce the notation*

$$E[\Gamma; \Omega]_n = \frac{n!}{(n-v)!} \prod_{i=1}^{d/2} \frac{(n-a_i)!}{n!},$$

where $v = |V_\Gamma \cup V_\Omega|$ and a_i is the number of π_i labels occurring in $\Gamma \cup \Omega$.

Given $(w; \vec{t}\,)$ and Ω, σ as above, we set $[\vec{t}; \sigma]$ to be the set of all pairs, $(\vec{s}, \tau)$, such that there is a permutation of the integers taking $\vec{t}$ to $\vec{s}$ and σ to τ; we set $[\vec{t}; \sigma]_n$ to the same, with the additional requirement that the components of $\vec{s}$ and

9. SELECTIVE TRACES IN GRAPHS WITH (WITHOUT) TANGLES

the image of τ lie in $\{1, \ldots, n\}$. If we let Γ be the form of w, sharing vertices and edges with Ω where σ and $\vec{t}$ "overlap," then

$$\mathrm{E}\left[\Gamma; \Omega\right]_n = \sum_{(\vec{s}, \tau) \in [\vec{t}; \sigma]_n} \mathrm{E}\left[\chi_{(w; \vec{s})} \chi_{\Omega; \tau}\right].$$

It follows that the expression in equation (39) is just

$$(40) \qquad \sum_{\Gamma \in \mathcal{W}_{<r}} W_\Gamma(\mathcal{W}, k) \mathrm{E}\left[\Gamma; \Omega\right]_n,$$

summed over one Γ from each Ω-isomorphism class (compare equation (24)).

We now go through the rest of Chapters 5, 6, and 8, indicating how to modify our results to allow deal with sums in equation (40). Let us begin by remarking that the expansion of Theorem 5.5 generalizes easily here, in that

$$\mathrm{E}\left[\Gamma; \Omega\right]_n = n^{v-e}\left(p_0 + \frac{p_1}{n} + \cdots + \frac{p_q}{n^q} + \frac{\text{error}}{n^{q+1}}\right)$$

where the $p_i = p_i(a_1, \ldots, a_{d/2}, v)$ are the expansion polynomials, and

$$|\text{error}| \le e^{qk/(n-k)} \bigl(v(v-1)/2 + a_1(a_1-1)/2 + \cdots + a_{d/2}(a_{d/2}-1)/2\bigr)^q$$

(by equation (20)). Since our potential walk is of length k, we have

$$v \le k + |V_\Omega|,$$

and

$$\sum a_i \le k + |E_\Omega|.$$

Since Ω is fixed, it follows that the error is bounded by ck^{2r-2} in the range $k \le n/2$, since the order of $\Gamma \cup \Omega$ is at least 1 (since $\Gamma \cup \Omega$ is pruned and contains an element of Ψ).

We next claim that given w (and our fixed Ω), the number of equivalence classes $[\vec{t}; \sigma]$ corresponding to w and Ω is bounded by $Ck^{2r+|V_\Omega|}$; indeed, Lemma 5.8 shows there at most Ck^{2r} $\vec{t}$ classes. The additional $|V_\Omega|$ choices of σ values can be chosen from at most $k + |V_\Omega| - 1$ "old" values plus one "new" value each time, for a total number of equivalence classes of at most Ck^{2r} times $(k + |V_\Omega|)^{|V_\Omega|}$, which is the claimed bound.

We obtain that the sum in equation (40) is

$$(41) \qquad \sum_{\Gamma \in \mathcal{W}_{<r}} W_\Gamma(\mathcal{W}, k) \left[\frac{p_0(\Gamma; \Omega)}{n^{\text{ord}(\Gamma; \Omega)}} + \cdots + \frac{p_{r-1-\text{ord}(\Gamma;\Omega)}(\Gamma; \Omega)}{n^{r-1}} + \frac{\text{error}(\Gamma; \Omega)}{n^r}\right]$$

with $p_i(\Gamma) = p_i\bigl(a_1(\Gamma; \Omega), \ldots, v(\Gamma; \Omega)\bigr)$ the expansion polynomials, with

$$\sum_{\Gamma \in \mathcal{W}_{<r}} W_\Gamma(\mathcal{W}, k) \text{error}(\Gamma; \Omega) = O(k^{4r-2+|V_\Omega|})(d-1)^k.$$

To attack the expansion polynomial sums in equation (41), we introduce straightforward generalizations of types and new types.

DEFINITION 9.8. *An Ω-type is an oriented graph $G_T = (V_T, E_T)$, with vertex and edge partial numberings such that (1) Ω is a subgraph of G_T, (2) all vertices of G_T except possibly the first one and possibly V_Ω vertices are of degree at least 3, and (3) all vertices and edges not in Ω are numbered.*

To a form, Γ, we associate its Ω-type, T, by taking $\Gamma \cup \Omega$ and suppressing the degree 2 vertices in $V_\Gamma \setminus V_\Omega$. T's edges are partially numbered since we number the edges in they order they must be traversed by a corresponding walk, and we don't number an edge if the walk doesn't traverse the edge (such edges are the edges of $E_\Omega \setminus E_\Gamma$).

DEFINITION 9.9. *A B-new Ω-type is a collection* $\widetilde{T} = (T; E_{\text{long}}, E_{\text{fixed}}; \vec{k}^{\text{fixed}})$, *(1) a lettered Ω-type, T, (2) a partition of E_T into two sets, $E_{\text{long}}, E_{\text{fixed}}$, (3) for each $e_i \in E_{\text{fixed}}$ an edge length, k_i^{fixed}, with $0 < k_i^{\text{fixed}} < B$, and (4) a Π^+-labelling of E_{fixed} with each $e_i \in E_{\text{fixed}}$ labelled with a word of length k_i^{fixed}. Furthermore, we require that E_Ω is contained in E_{fixed}, and that each E_Ω edge has length 1 in $\widetilde{T}$ and is labelled with its Ω label in $\widetilde{T}$.*

So conditions (2)–(4) are just as for a new type (See Definition 5.19), and condition (1) here involves an Ω-type instead of a type.

Again, it is easily seen that there are finitely many (isomorphism classes) of Ω-types of a given order, and finitely many B-new Ω-types belonging to a type, T, for a given B (and T). It suffices to show that for B and S sufficiently large with $B > S$ we have that for any B-new Ω-type, $\widetilde{T}$, and any polynomial, p, in the a_i's we have
$$\sum_{\Gamma \in \widetilde{T}} W_\Gamma(\mathcal{W}, k) p\big(a_1(\Gamma; \Omega), \ldots, a_{d/2}(\Gamma; \Omega)\big)$$
is d-Ramanujan.

For integers $\{m_i\}$ indexed over e_i edges of E_T, set
$$W_{\widetilde{T}}(\vec{m}; S, \Psi')$$
to the number of walk classes of new type $\widetilde{T}$, traversing the edge e_i m_i times, that are irreducible (S, Ψ') selective cycles and that respect the T partial numbering and orientation; we assume $m_i \geq 1$ or $m_i = 0$ according to whether or not e_i is numbered (i.e., e_i is to be traversed on our walk); assuming $\widetilde{T}$ is a B-new type with $B > S$, this number does not depend on the "long" (i.e., E_{long}) edge lengths. Index the edges so that $E_{\text{long}} = \{e_1, \ldots, e_t\}$ and $E_{\text{fixed}} = \{e_{t+1}, \ldots, e_{b'}\}$, with $\{e_{t+1}, \ldots, e_b\}$ the numbered edges (so $m_{b+1} = \cdots = m_{b'} = 0$). We claim that the proof of Theorem 6.6 shows that
$$W_{\widetilde{T}, S}(M_1, M_2) = \sum_{\substack{m_1 + \ldots + m_t = M_1 \\ m_{t+1} + \cdots + m_b = M_2}} W_{\widetilde{T}}(\vec{m}; S, \Psi'),$$
satisfies the bound
$$W_{\widetilde{T}, S}(M_1, M_2) \leq cB(\sqrt{d-1} - \epsilon)^{cM_1 + M_2}.$$

for $S \geq S_0$ for some c, S_0, and ϵ depending only on $\widetilde{T}$. This is because the argument of Theorem 6.6 is unaffected by the two essential new features that T has over types, which are the possible presence of (1) some more degree 2 vertices (other than just the vertex numbered 1), and (2) some edges whose lengths are fixed at 1 (namely E_Ω edges). But each $a_i = a_i(\Gamma; \Omega)$, which is the number of edges labelled π_i in $\Gamma \cup \Omega$, has
$$a_i(\Gamma; \Omega) = a_i(\Gamma) + a_i(\Omega \setminus \Gamma),$$

and $a_i(\Omega \setminus \Gamma)$ is fixed in a new Ω-type, $\widetilde{T}$ (by the edge partial numbering). Thus for any polynomial p and new typoid, $\widetilde{T}$, there is a polynomial $\widetilde{p}$ such that

(42) $$p\big(a_1(\Gamma;\Omega),\ldots,a_{d/2}(\Gamma;\Omega)\big) = \widetilde{p}\big(a_1(\Gamma),\ldots,a_{d/2}(\Gamma)\big).$$

But it is easy to see that Theorem 8.2 holds if we extend the definition of type and new type to allow any fixed number of vertices of degree 2. Furthermore the condition that Γ belongs to $\widetilde{T}$ is the same as Γ belonging to this extended notion of new type. We conclude that either side of equation (42) is d-Ramanujan. □

CHAPTER 10

Strongly Irreducible Traces

We wish to use Theorem 9.3 to estimate eigenvalues. However, it is easier to use *strongly irreducible traces* rather than irreducible traces. We explain this and develop the properties of the strongly irreducible trace in this section.

DEFINITION 10.1. *A word $w \in \Pi^*$ is* strongly irreducible *if w is irreducible and $w = \sigma_1 \ldots \sigma_k$ with $\sigma_1 \neq \sigma_k^{-1}$.*

For any of our irreducible traces, selective irreducible traces, irreducible walk sums, etc., we can form its "strongly irreducible" version where we discard contributions from words that are not strongly irreducible. In any graph, labelled or not, one can speak of strongly irreducible closed walks as those closed walks that are irreducible and whose last step is not the opposite of its first step.

DEFINITION 10.2. *The k-th* strongly irreducible trace *of a graph, G, is the number of strongly irreducible closed walks of length k for a positive integer k; we denote it $\mathrm{SIT}(G, k)$ or $\mathrm{SIT}(A, k)$ if A is the adjacency matrix of G. If G has half-loops, we consider each half-loop to be a strongly irreducible closed walk (of length 1) and include it in our count for $\mathrm{SIT}(G, 1)$ or $\mathrm{SIT}(A, 1)$.*

Half-loops are only a concern for us in the model $\mathcal{J}_{n,d}$; the reason that we count half-loops as strongly irreducible is to make Lemma 10.4 hold.

With a closed walk of length k in G_{Irred} about a directed edge, e, of G, we may associate the strongly irreducible closed walk in G about the vertex in which e originates. It follows that if μ_i are the eigenvalues of G_{Irred}, we have

$$(43) \qquad \mathrm{SIT}(G,k) = \sum_{i=1}^{nd} \mu_i^k$$

for all $k \geq 2$; for $k = 1$ we must add the number of half-loops to the right-hand-side of the above equation, since there is no edge in G_{Irred} from a vertex to itself when the vertex corresponds to a half-loop.

We will study the relationship between $\mathrm{IrredTr}\,(G,k)$ and $\mathrm{SIT}(G,k)$ and its consequences. The most important consequence is the following theorem.

THEOREM 10.3. *For $|\lambda| \leq d$, let*

$$\mu_{1,2}(\lambda) = \frac{\lambda \pm \sqrt{\lambda^2 - 4(d-1)}}{2},$$

and set

$$(44) \qquad \widetilde{q}_k(\lambda) = \mu_1^k(\lambda) + \mu_2^k(\lambda) + \left(1 + (-1)^k\right)(d-2)/2.$$

If G is a d-regular graph with no half-loops and adjacency matrix eigenvalues $\lambda_1 \geq \cdots \geq \lambda_n$, then

$$\text{SIT}(G,k) = \sum_{i=1}^{n} \widetilde{q}_k(\lambda_i). \tag{45}$$

Furthermore, if instead G has no whole-loops, then the same is true with $\widetilde{q}_k$ replaced by $\widehat{q}_k$ where

$$\widehat{q}_k(x) = \begin{cases} \widetilde{q}_k(x) & \text{if } k \text{ is even or } k = 1, \\ \widetilde{q}_k(x) - x & \text{if } k \geq 3 \text{ is odd}. \end{cases}$$

Furthermore, we shall see that $\widetilde{q}_k$, like the q_k of Lemma 2.3, are polynomials of degree k that may alternatively be expressed as a simple linear combination of Chebyshev polynomials (plus the ± 1 eigenvalue contribution for the $\widetilde{q}_k$). (In a sense, equation (45) says that to each eigenvalue, λ, of G, there correspond eigenvalues $\mu_{1,2}(\lambda)$ of multiplicity one each and eigenvalues 1 and -1 of multiplicity $(d-2)/2$ each in G_{Irred}.)

LEMMA 10.4. *Let G be a d-regular graph on n vertices with h half-loops. Then for all integers $k \geq 2$ we have[1]*

$$\text{IrredTr}\,(G,k) = \text{SIT}(G,k) + (d-2)\sum_{i=1}^{\lfloor (k-1)/2 \rfloor} (d-1)^{i-1}\text{SIT}(G,k-2i)$$

$$+ \begin{cases} 0 & \text{if } k \text{ is even,} \\ (d-1)^{(k-3)/2}h & \text{if } k \text{ is odd.} \end{cases}$$

Therefore for all $k \geq 4$ (and for $k = 3$ when $h = 0$) we have

$$\text{IrredTr}\,(G,k) - (d-1)\text{IrredTr}\,(G,k-2) = \text{SIT}(G,k) - \text{SIT}(G,k-2) \tag{46}$$

PROOF. The last equation follows from the previous one, with a little care when $k = 3$; indeed, for $k = 3$ we have $\text{SIT}(G,k-2) = \text{SIT}(G,1) = 2w + h$, where w, h are the number of whole- and half-loops. So

$$\text{IrredTr}\,(G,3) = \text{SIT}(G,3) + (d-2)\text{SIT}(G,1) + h,$$

and

$$\text{IrredTr}\,(G,1) = \text{SIT}(G,1) = \text{Trace}\,(A) = 2w + h.$$

Hence

$$\text{IrredTr}\,(G,3) - (d-1)\text{IrredTr}\,(G,1) - \text{SIT}(G,3) + \text{SIT}(G,1) = h.$$

Thus equation (46) holds with $k = 3$ if $h = 0$. We similarly show that this equation holds regardless of h for $k \geq 4$.

So it suffices to prove the first equation of the lemma. Each irreducible closed walk about a vertex, v, begins by traversing a path, p, to a vertex, w, then follows a strongly irreducible (nonempty) closed walk about w, and then backtracks over p (and this statement is only true if we count half-loops as strongly irreducible); each irreducible closed walk has a uniquely determined such p and w. So we may count irreducible closed walks of length k in G by counting how many paths of length i there are from w that when combined with a strongly irreducible closed walk about w of length $k - 2i$ yield an irreducible closed walk, C, of length k. The strongly

---

[1]For $k = 2$ the summation in the formula to follow is ignored, since it ranges from $i = 1$ to $i = \lfloor (k-1)/2 \rfloor = 0$.

irreducible closed walk's length, $k - 2i$, must be positive, or else C isn't irreducible. If C is of length at least 2 or is a whole-loop, then there are $d - 2$ possibilities for the first edge of the path (since two edges are ruled out by C in either case), and $d - 1$ possibilities for all edge choices thereafter. For a half-loop there are $d - 1$ possibilities for the first edge (since only the single half-loop edge is ruled out). So the contribution per strongly irreducible closed walk of length $k - 2i$ is 1 for $i = 0$, $(d-2)(d-1)^{i-1}$ for $i \geq 1$, except in a half-loop, where the contribution per half-loop is $(d-1)^i$, i.e., an additional $(d-1)^{i-1}$ beyond the standard contribution. Since k is odd and $i = (k-1)/2$ in the case of a half-loop, the additional amount beyond the standard contribution is $(d-1)^{(k-3)/2}$ per half-loop. $\square$

We return to the proof of the theorem. First assume that G has no half-loops. We will prove by induction on $k \geq 1$ that

$$\text{SIT}(G, k) = \sum_{i=1}^{n} \widetilde{q}_k(\lambda_i), \tag{47}$$

where $\widetilde{q}_k$ are polynomials of degree k. Clearly

$$\text{SIT}(G, 1) = \text{Trace}(A),$$

and so $\widetilde{q}_1$ exists as desired with $\widetilde{q}_1(\lambda) = \lambda$. Of the closed walks of length 2, all irreducible closed walks are strongly irreducible, so $\widetilde{q}_2(\lambda) = \lambda^2 - d$. Lemmas 10.4 and 2.3 now imply (by induction on k) that polynomials $\widetilde{q}_k(\lambda)$ exist of degree k satisfying equation (47), and that the $\widetilde{q}_k$, for $k \geq 2$, are annihilated by

$$(\sigma_k^2 - 1)(\sigma_k^2 - \lambda \sigma_k + (d-1)), \tag{48}$$

where σ_k is the "shift in k" operator, i.e., $\sigma_k(f(k)) = f(k+1)$ (here we use the fact that the q_k are annihilated by $\sigma_k^2 - \lambda \sigma_k + (d-1)$, mentioned below Lemma 2.3). Since $\mu_{1,2}(\lambda), \pm 1$ are the four roots in σ_k of equation (48), we have

$$\widetilde{q}_k(\lambda) = c_1 \mu_1^k + c_2 \mu_2^k + c_3 + c_4(-1)^k,$$

where the $c_i = c_i(\lambda)$ assuming that the four roots are distinct. There are now two ways to finish the theorem.

The first method to finish the theorem is to calculate $\widetilde{q}_3, \widetilde{q}_4$ and verify equation (44) holds for $k = 2, 3, 4, 5$, i.e., that $c_1 = c_2 = 1$ and $c_3 = c_4 = (d-2)/2$ work in those cases. Then by uniqueness (i.e., the nonvanishing of a Vandermonde determinant), those c_i's must be the unique c_i's that work for all k.

Another way to finish the theorem is to use the fact that $c_1(d) = 1$ (see the remark after Lemma 5.10), and then argue that $c_1(\lambda) = c_2(\lambda) = 1$ by analytic continuation. First, the c_i's are the unique solutions to a 4×4 system of equations with coefficient analytic in λ; hence the c_i are, indeed, analytic in λ. Next, notice that $\mu_1(\lambda)$ at $\lambda = d$ analytically continues to $\mu_2(\lambda)$ at $\lambda = d$ by one loop about $2\sqrt{d-1}$; thus is suffices to prove that $c_1 = 1$ near d. Next notice that different λ's give different μ's (indeed, $r^2 - \lambda_1 r + (d-1) = 0$ and $r^2 - \lambda_2 r + (d-1) = 0$ for the same r implies $r(\lambda_2 - \lambda_1) = 0$), so if $\lambda \neq d$ is an eigenvalue of multiplicity k in a graph, then $c_1(\lambda)$ times k must be an integer. But there is a sequence, $z_n \to 1$, with $c_1(z_n)$ an integer or half-integer (namely a cycle of length n where each edge has multiplicity $d/2$ has $z_n = (d/2)\cos(2\pi/n)$ as an eigenvalue of multiplicity two). So by continuity $c_1(z_n) = 1$ for sufficiently large n, and thus c_1 is identically 1.

It suffices to determine c_3, c_4, which from $\widetilde{q}_1, \widetilde{q}_2$ we find are (the constant functions) $c_3 = c_4 = (d-2)/2$.

Finally, if G has only half-loops (no whole-loops), then the k even formula and $k = 1$ formula are unchanged. We easily see by induction on odd $k \geq 3$ that the polynomial $\widehat{q}_k = \widetilde{q}_k(x) - x$ works (we use the fact that h, the number of half-loops, is the trace of A_G, i.e., $h = \sum \lambda_i$). □

THEOREM 10.5. *Fix an integer $d > 2$ and a real $\epsilon > 0$. There is an $\eta > 0$ such that if G is a d-regular graph with $|\lambda_i| \leq d - \epsilon$ for all $i > 1$, then the μ_i of equation (43) satisfy $\mu_1 = d - 1$ and $|\mu_i| \leq d - \eta$ for all $i > 1$.*

PROOF. The μ_i must be ± 1 or roots of the equation in μ
$$\mu^2 - \mu\lambda_i + (d-1) = 0,$$
or
$$\mu = \frac{\lambda_i \pm \sqrt{\lambda_i^2 - 4(d-1)}}{2}.$$
For $i = 1$ we have $\lambda_1 = d$ and the corresponding μ's are $\mu = d - 1, 1$. The other μ's are either 1 or come from λ_i with $i > 1$. But for $|\lambda_i| \leq d - \epsilon$ it is easy to see that the corresponding μ's are bounded away from $d - 1$. □

We can form a selective, strongly irreducible trace by taking
$$\text{SSIT}_{S,\Psi'}(G;k)$$
to be the number of strongly irreducible closed walks of length k that are (S, Ψ')-selective. Notice that there are no more strongly irreducible closed walks than irreducible closed walks in any lettered type, and the strong irreducibility of a potential walk can be determined from its image in the corresponding lettered type. Hence the expansion theorems of Chapters 6–9, especially Theorem 9.3, carry over to SSIT replacing IrSelTr, by simply replacing
$$W_{\widetilde{T},S}(M_1, M_2)$$
by the same number of walk classes of the new Ω-type, $\widetilde{T}$, except requiring that the walks are strongly irreducible.

CHAPTER 11

A Sidestepping Lemma

LEMMA 11.1. *Fix integers $r, \widetilde{r}, d$ with $d > 2$, polynomials $p_0, \ldots, p_r$, a constant, c, and an integer D. Assume that for each n we have complex-valued random variables $\theta_1, \ldots, \theta_m$ such that $m = Dn$. Assume that $1 - \theta_i$ is of absolute value at most 1, and is purely real if its absolute value is greater than $(d-1)^{-1/2}$. Furthermore assume that for all integers $k \geq 1$ we have*

$$(49) \qquad E\left[\sum_i (1-\theta_i)^k\right] = \sum_{j=0}^{r-1} p_j(k) n^{-j} + O\bigl(k^{\widetilde{r}} n^{-r} + k^c (d-1)^{-k/2}\bigr).$$

Then for sufficiently large n we have

$$(50) \qquad E\left[\sum_i \chi_{\{|\theta_i| > \log^{-2} n\}} (1-\theta_i)^k\right] = O\bigl(Dn^{1-(r/3)} + k^c(d-1)^{-k/2}\bigr)$$

for all k with $1 \leq k \leq n^\gamma$ for some constant $\gamma > 0$, where the constant γ and the constant in the $O(\,\cdot\,)$ notation depends only on $r, \widetilde{r}, d$ and the maximum degree of the p_i.

After this section we will apply this lemma with the $(d-1)(1-\theta_i)$ being the eigenvalues of G_{Irred}, with k proportional (for fixed r) to $\log n$.

If σ_k denotes the "shift with respect to k," i.e., $\sigma_k\bigl(f(k)\bigr) = f(k+1)$, then some fixed power of $\sigma_k - 1$ annihilates the $p_j(k)$, and also

$$(\sigma_k - 1)^i (1-\theta)^k = (-\theta)^i (1-\theta)^k.$$

This allows us to say a lot about the θ_i by applying some power of $\sigma_k - 1$ to equation (49). For the irreducible trace, the $(1 - \theta_i)^k$ are replaced by certain Chebyshev polynomials of θ_i, and applying powers of $\sigma_k - 1$ to them seems more awkward; this is why we have introduced strongly irreducible traces.

In [**Fri91**], we worked with irreducible traces, not strongly irreducible traces. There we had the $(1-\theta_i)^k$ replaced by Chebyshev polynomials of $1-\theta_i$; however (1) we knew that $\theta_1 = 0$ and θ_i was bounded away from 0 with probability $1-O(n^{1-d})$, and (2) we could only prove the asymptotic expansion up to r which was roughly proportional to $d^{1/2}$. So we could directly apply the analogue of equation (49) with k roughly $\log^2 n$ to determine that $p_0(k) = 1$ and the higher p_j vanish (up to j roughly proportional to $d^{1/2}$). In this paper the arbitrary length of the asymptotic expansion for a type of trace comes at the cost of having far less control over the θ_i, and we have no ability to determine the p_j exactly. Fortunately Lemma 11.1 allows us to control the θ_i bounded away from 0, and fortunately we will see that the θ_i for $i > 1$ are bounded away from 0 with probability $1 - O(n^{-\tau_{\text{fund}}})$.

We wish to comment that one expects polynomials $p_j = p_j(k)$ to arise from the binomial expansion. Namely (by Taylor's theorem),

$$(1-\theta)^k = \sum_{i=0}^{s-1} \binom{k}{i}(-\theta)^i + \binom{k}{s}(-\theta)^s(1-\xi)^{k-s}, \tag{51}$$

for some $\xi \in [0, \theta]$. So for those $\theta \leq n^{-\beta}$ for some constant $\beta > 0$, we may take s roughly r/β and get an error term in ξ bounded by roughly $O(n^{-r})$; we get a similarly bounded error term when taking expected values of $(1-\theta)^k \chi_E$ where E is the event that $\theta \leq n^{-\beta}$. In this way, equation (51) could give rise to the terms of an asymptotic expansion.

PROOF. (of Lemma 11.1). Let s be a fixed even integer such that the maximum degree of the p_j is at most $s-1$. We apply $(\sigma_k - 1)^s$ to equation (49) to conclude that

$$E\left[\sum_i (-\theta_i)^s(1-\theta_i)^k\right] = O(k^{\tilde{r}} n^{-r} + k^c (d-1)^{-k/2})$$

for any k, where the constant in the $O(\,\cdot\,)$ notation depends on d and s. We conclude that for $\log^2 n \leq k \leq n^{r/(2\tilde{r})}$ we have

$$E\left[\sum_i \theta_i^s (1-\theta_i)^k\right] = O(n^{-r/2}).$$

Applying this for $k = \lfloor n^\gamma \rfloor$ and $k = \lfloor \log^2 n \rfloor$, where $\gamma = r/(2\tilde{r})$, and subtracting we conclude

$$E\left[\sum_i \theta_i^s\left((1-\theta_i)^{\lfloor \log^2 n \rfloor} - (1-\theta_i)^{\lfloor n^\gamma \rfloor}\right)\right] = O(n^{-r/2}).$$

Since θ_i is real unless $1 - \theta_i = (d-1)^{-1/2}$, we conclude that

$$E\left[\sum_i |\theta_i|^s \left(|1-\theta_i|^{\lfloor \log^2 n \rfloor} - |1-\theta_i|^{\lfloor n^\gamma \rfloor}\right)\right]$$

$$= O(n^{-r/2}) + O\left(n(d-1)^{-(\log^2 n)/2}\right) = O(n^{-r/2}).$$

Now for any $\theta_i \leq \log^{-2} n$ we have $(1-\theta_i)^{\log^2 n}$ is at least roughly $1/e$ for large n; also for $\theta_i \geq n^{-\alpha}$ for a constant $\alpha > 0$ we have $(1-\theta_i)^{n^\gamma}$ is near 0 for large n provided $\alpha < \gamma$, and also $\theta_i^s \geq n^{-s\alpha}$. We conclude that

$$E\left[\sum_i (\chi_{\{n^{-\alpha} \leq \theta_i \leq \log^{-2} n\}}) n^{-s\alpha}\right] = O(n^{-r/2}),$$

and hence, since θ_i is real for $|\theta_i| < \log^{-2} n$ for n large,

$$E\left[\chi_{n^{-\alpha} \leq \theta_i \leq \log^{-2} n}\right] = O(n^{s\alpha - (r/2)}). \tag{52}$$

Let $\alpha > 0$ be fixed with $s\alpha < r/6$, and set

$$\begin{aligned} A_1[i,n] &= \text{The event that } \theta_i < n^{-\alpha}, \\ A_2[i,n] &= \text{The event that } n^{-\alpha} \leq \theta_i \leq \log^{-2} n, \\ A_3[i,n] &= \text{The event that } \log^{-2} n < |\theta_i|. \end{aligned}$$

Equation (52) implies that
$$\text{Prob}\{A_2[i,n]\} = O(n^{-r/3}).$$
Since the number of θ_i is linear in n, we conclude that

(53) $$\text{E}\left[\sum_i \chi_{A_2[i,n]}(1-\theta_i)^k\right] = O(n^{1-(r/3)}).$$

By the comment just before the proof, we have (using Taylor's theorem)
$$\text{E}\left[\chi_{A_1[i,n]}(1-\theta_i)^k\right] = \sum_{j=0}^{j\leq 2r/\alpha} \binom{k}{j}\text{E}\left[\chi_{A_1[i,n]}(-\theta_i)^j\right] + O(k^{1+(2r/\alpha)}n^{-2r}).$$

Summing over i in the above involves summing over i in the expected values and multiplying the error term by a number linear in n. So let
$$q(k,n) = \sum_i \sum_{j=0}^{j\leq r/\alpha} \binom{k}{j}\text{E}\left[\chi_{A_1[i,n]}(-\theta_i)^j\right],$$
which is a polynomial of fixed degree in k whose coefficients depend on n (and the θ_i which are given for each value of n). Fix a γ for which
$$O(k^{1+(2r/\alpha)}n^{1-2r}) = O(n^{-r/3})$$
for all $k \leq n^\gamma$, i.e., fix a γ with
$$\gamma(1+(r/\alpha)) \leq 5r/3 - 1.$$
Then for all $k \leq n^\gamma$ we have

(54) $$\text{E}\left[\sum_i \chi_{A_1[i,n]}(1-\theta_i)^k\right] = q(k,n) + O(n^{-r/3}).$$

Now combine
$$\text{E}\left[\sum_i (1-\theta_i)^k\right] = \sum_{j=1}^{3} \text{E}\left[\sum_i \chi_{A_j[i,n]}(1-\theta_i)^k\right]$$
with equations (54) and (53) to conclude that for $k \leq n^\gamma$ we have
$$\text{E}\left[\sum_i (1-\theta_i)^k\right] = q(k,n) + \text{E}\left[\sum_i \chi_{A_3[i,n]}(1-\theta_i)^k\right] + O(n^{1-(r/3)}).$$
On the other hand, equation (49) just says
$$\text{E}\left[\sum_i (1-\theta_i)^k\right] = p(k,n) + O\left(k^{\tilde{r}}n^{-r} + k^c(d-1)^{-k/2}\right),$$
where $p(k,n)$ is the polynomial in k given as the sum of the $p_j(k)/n^j$. Therefore
$$\text{E}\left[\sum_i \chi_{A_3[i,n]}(1-\theta_i)^k\right]$$

(55) $$= p(k,n) - q(k,n) + O(n^{1-(r/3)}) + O\left(k^{\tilde{r}}n^{-r} + k^c(d-1)^{-k/2}\right)$$

for all $k \leq n^\gamma$. Since $A_3[i,n]$ implies $|1-\theta_i| \leq (1-\log^{-2} n)$ and thus $|1-\theta_i|^k \leq e^{O(\log^{-2} n)}$ for $k \geq \log^4 n$, we have

$$\text{E}\left[\sum_i \chi_{A_3[i,n]}|1-\theta_i|^k\right] = O(n^{-r/3}) \tag{56}$$

for $k \geq \log^4 n$ for n sufficiently large. We conclude that

$$p(k,n) - q(k,n) = O(n^{1-r/3}) \tag{57}$$

for all k with $\log^4 n \leq k \leq n^\gamma$.

SUBLEMMA 11.2. *Let $g(k)$ be a polynomial in k of degree $\leq s-1$ such that $|g(i)| \leq 1$ for integers $i = a, a+1, \ldots, b$ for some integers a, b with $a \leq b$. Then $|g(i)| \leq 2^s - 1$ for integers i with*

$$a - \frac{b-a}{s-1} \leq i \leq a.$$

PROOF. We have $(\sigma_k - 1)^s g = 0$, and therefore

$$g(x) = \sum_{i=1}^{d} \binom{s}{i}(-1)^{i-1} g(x+hi)$$

for any x and h. Given $i < a$, let $h = a - i$ and $x = i$ in the above; $x+h, x+2h, \ldots, x+sh$ are integers between a and b provided that

$$i + (a-i)s \leq b,$$

so $i \geq (as-b)/(s-1)$ or $i \geq a - (b-a)/(s-1)$. If so, then

$$|g(i)| \leq \sum_{i=1}^{d} \binom{s}{i} |g(x+hi)| \leq \sum_{i=1}^{d} \binom{s}{i} = 2^s - 1.$$

$\square$

Recall equation (57) and the fact that p and q are polynomials in k (for fixed n) of bounded degree. So applying the above sublemma for $a = 2\lceil(\log^4 n)/2\rceil$ and $b = 2\lfloor n^\gamma/2 \rfloor$ implies that $p(k,n) - q(k,n) = O(n^{1-(r/3)})$ for $1 \leq k \leq \log^4 n$. Equation (57) now holds for all $1 \leq k \leq n^\gamma$. We conclude that equation (56) holds for all $1 \leq k \leq n^\gamma$. Adding this to equation (53) yields the desired equation (50). $\square$

CHAPTER 12

Magnification Theorems

In this section we use standard counting arguments to prove theorems implying "magnification" or "expansion" for "most" random graphs; here "most" means all graphs excepting a set of probability $O(n^{-\tau_{\text{fund}}})$. These theorems will then be used with Lemma 11.1 to prove Theorems 1.1, 1.2, and 1.3.

A graph, G, with n vertices is said to be a γ-magnifier if for all subsets of vertices, A, of size at most $n/2$ we have

$$|\Gamma(A) - A| \geq \gamma |A|,$$

where $\Gamma(A)$ denotes those vertices connected to some member of A by an edge. Alon has shown that any d regular γ-magnifier has

$$\lambda_2(G) \leq d - \frac{\gamma^2}{4 + 2\gamma^2}$$

(see [**Alo86**]; see [**Dod84, SJ89, JS89**] for related "edge magnification" results).

DEFINITION 12.1. *Say that a d-regular graph on n vertices is a γ-spreader if for every subset, A, of at most $n/2$ vertices we have*

$$|\Gamma(A)| \geq (1 + \gamma)|A|.$$

THEOREM 12.2. *Let G be a d-regular γ-spreader. Then for all $i > 1$ we have*

$$\lambda_i^2(G) \leq d^2 - \frac{\gamma^2}{4 + 2\gamma^2}.$$

PROOF. Since the graph is d-regular, we have $|\Gamma(B)| \geq |B|$ for all subsets of vertices, B. Taking $B = \Gamma(A)$ yields

$$|\Gamma^2(A)| \geq |\Gamma(A)| \geq (1 + \gamma)|A|.$$

Hence G^2, the graph on V_G whose edges are paths in G of length 2 (and whose adjacency matrix is A_G^2), is a d^2-regular γ-magnifier. Now apply Alon's result on magnification and eigenvalues to G^2, whose eigenvalues are $\lambda_i^2(G)$. □

We now establish that for all our models, a graph will be a γ-spreader for some fixed $\gamma = \gamma(d) > 0$ with probability $1 - O(n^{-\tau_{\text{fund}}})$.

THEOREM 12.3. *For any $\epsilon > 0$ and even $d \geq 4$ there is a $\gamma > 0$ such that $G \in \mathcal{G}_{n,d}$ is a γ-spreader with probability $1 - O(n^{-\tau_{\text{fund}}})$.*

Later we shall prove this theorem for other models of random graphs, by very similar calculations. This theorem is easy for d sufficiently large; but when $d = 4$ (or later possibly $d = 3$) one has to calculate fairly carefully.

PROOF. Fix $A, B \subset \{1, \ldots, n\}$, and consider the event that $\Gamma(A) \subset B$. We will impose the condition that $a = |A| \leq n/2$ and $|B| = a + \lfloor \gamma a \rfloor$.

Fix constants, γ, C with $0 < \gamma < 1/C$. Consider the situation where $|A| < C$. In this case $|B| = |A| = a$. But since G is d-regular and we have $d|A|$ edges leaving A, these edges comprise all edges incident upon B (since $|B| = |A|$). Thus $A \cup B$ is a union of connected components of G. But this cannot occur if G has no supercritical tangles of size at most $2C$ (since each connected component of G has $\lambda_{\text{Irred}} = d - 1$). For a constant C there are only a constant number of tangles of size at most $2C$. Thus, by forsaking a probability of $O(n^{-\tau_{\text{fund}}})$, we may assume that $a = |A| \geq C$ for any fixed constant, C (provided that we then take $\gamma < 1/C$) for sufficiently large n.

So consider a random permutation, $\pi = \pi_i$, and consider the event that π and π^{-1} map A to B. Let $C_1 = A \cap B$, $C_2 = A \setminus B$, $C_3 = B \setminus A$, and let $c_i = |C_i|$. We view π as determined by a perfect matching of a bipartite graph on inputs, I, and outputs, O, with I, O being copies of $\{1, \ldots, n\}$ (and $i \in I$ mapped to $\pi(i) \in O$). Viewing π as a bipartite matching, it consists of (1) r edges from C_1 to C_1 (i.e., the I vertices corresponding to C_1 to those O vertices corresponding to C_1), (2) $c_1 - r$ edges from C_1 to C_3, (3) $c_1 - r$ edges from C_3 to C_1, (4) c_2 edges from C_2 to C_3, and (5) c_2 edges from C_3 to C_2. (This is true since a C_2 vertex, either input or output, must be paired with a C_3 vertex, and a C_1 vertex must be paired with either a C_1 or C_3 vertex.) So the event that π and π^{-1} map A to B with c_1, c_2, c_3, r, A, B all held fixed has probability

$$p(c_1, c_2, c_3, r) = \left[\binom{c_1}{r}^2 r!\right] \left[\binom{c_3}{c_1 - r}(c_1 - r)!\right]^2 \times$$

$$\left[\binom{c_3 - c_1 + r}{c_2} c_2!\right]^2 [n(n-1) \cdots (n - 2c_1 - 2c_2 + r + 1)]^{-1}$$

(The first expression in square brackets corresponds to choosing r C_1 to C_1 edges; the second expression corresponds to choosing $c_1 - r$ C_1 to C_3 edges, and is squared to include choosing the C_3 to C_1 edges; etc.) The probability taken over all A, B of a given c_1, c_2, c_3 (and with r fixed) is therefore at most

$$(58) \qquad \binom{n}{c_1, c_2, c_3, n - c_1 - c_2 - c_3} p^{d/2}(c_1, c_2, c_3, r).$$

It suffices to show that this expression is $O(n^{-s})$ with $s = \tau_{\text{fund}} + 4$, provided that a is sufficiently large (and at most $n/2$), since then we can sum equation (58) over the at most n^4 relevant values of c_1, c_2, c_3, r. We should remind ourselves that c_1, c_2, c_3, r range over integers with

$$c_1 + c_2 = a, \quad c_1 + c_3 = a + \lfloor \gamma a \rfloor, \quad r \leq c_1.$$

Furthermore, considering the expression defining p, we have $c_3 - c_1 + r \geq c_2$.

We now write

$$(59) \qquad b = b(c_1, c_2, c_3, r, n) = \binom{n}{c_1, c_2, c_3, n - c_1 - c_2 - c_3}$$

$$= \frac{n!}{c_1! \, c_2! \, c_3! \, (n - c_1 - c_2 - c_3)!}$$

and

(60) $$p = p(c_1, c_2, c_3, r, n) = \frac{(c_1!\, c_3!)^2\, (n - 2c_1 - 2c_2 + r)!}{((c_1 - r)!\, (c_3 - c_1 - c_2 + r)!)^2\, r!\, n!}.$$

We make some general remarks about analyzing the factorials in the above two equations:

(1) All factorials in the above equations are of the form $(\mu n)!$ for some $\mu \in [0, 1]$. Stirling's formula $m! \sim (m/e)^m \sqrt{2\pi m}$ implies that

(61) $$\frac{1}{n} \log[(\mu n)!] = \mu \log(n/e) + \mu \log \mu + O\left(\frac{\log n}{n}\right),$$

where the constant in the $O(\,\cdot\,)$ is independent of n and $\mu \in [0, 1]$.

(2) In analyzing b and p above, we may ignore the $\mu \log(n/e)$ term in equation (61). This is because b, p are *balanced* in that the sum of the numbers to which factorials are applied is the same in the numerator and denominator; in other words, the $\mu \log(n/e)$ terms in the numerator will exactly cancel those in the denominator.

(3) Let $f(\theta) = -\theta \log \theta$. We claim that for $\theta_1, \theta_2 \in [0, 1]$ we have
$$|f(\theta_1) - f(\theta_2)| \leq \max\bigl(f(|\Delta\theta|), f(1 - |\Delta\theta|)\bigr), \qquad \text{with} \quad \Delta\theta = \theta_2 - \theta_1.$$

Indeed, since $f''(\theta) = -1/\theta < 0$ for $\theta > 0$, f is concave in $[0, 1]$, and so $g(\theta) = f(\theta + \Delta\theta) - f(\theta)$ is decreasing in θ for $\Delta\theta$ fixed; so $|g|$'s maximum over an interval is taken at its endpoints, and since $f(0) = f(1) = 0$, the above claim is established.

Next, a Taylor expansion shows that $-\epsilon \log \epsilon \geq -(1 - \epsilon) \log(1 - \epsilon)$ for sufficiently small $\epsilon > 0$. Hence there is an ϵ_0 such that

(62) $$|f(\theta_1) - f(\theta_2)| \leq f(|\theta_1 - \theta_2|)$$

for all $\theta_1, \theta_2 \in [0, 1]$ with $|\theta_1 - \theta_2| \leq \epsilon_0$.

Let $\nu_i, \rho, \alpha, \delta$ ($i = 1, 2, 3$) be the non-negative reals given by
$$c_i = \nu_i n, \qquad r = \rho n, \qquad a = \alpha n, \qquad \lfloor \gamma a \rfloor = \delta n.$$

We have that
$$\nu_1 + \nu_2 = \alpha, \qquad \nu_1 + \nu_3 = \alpha + \delta, \qquad \rho \leq \nu_1, \qquad \rho \geq \nu_1 + \nu_2 - \nu_3.$$

We conclude that
$$|\nu_2 - \nu_3| \leq \delta, \qquad |\nu_1 - \rho| \leq \delta.$$

It follows from equation (61), remark (2) below it, and equation (62), that we may replace ν_3 with ν_2 and ρ with ν_1 in calculating $(\log b)/n$ and incur an additive error term of at most $O(\delta \log \delta)$. Thus we get

$$\frac{\log b}{n} = h(\nu_1, \nu_2) + O\left(|\delta \log \delta| + \frac{\log n}{n}\right),$$

where

(63) $$h(\nu_1, \nu_2) = -\nu_1 \log \nu_1 - 2\nu_2 \log \nu_2 - (1 - \nu_1 - 2\nu_2) \log(1 - \nu_1 - 2\nu_2).$$

Similarly we calculate
$$\frac{-\log p}{n} = h(\nu_1, \nu_2) + O\left(|\delta \log \delta| + \frac{\log n}{n}\right),$$

i.e., we have the exact same equation (!) for $\log b$ replaced by $-\log p$ (this "coincidence" happens for the other models as well). Hence

$$\frac{\log(bp^2)}{n} = -h(\nu_1, \nu_2) + O\left(|\delta \log \delta| + \frac{\log n}{n}\right).$$

Since $\nu_1 + \nu_2 = \alpha$, we have either (or both) ν_i are $\geq \alpha/2$. Hence

$$h(\nu_1, \nu_2) \geq -(\alpha/2) \log(\alpha/2).$$

Now we claim that for any constant $C > 0$ there is a constant $\gamma > 0$ such that for all $\alpha \in [0, 1/2]$ we have

(64) $$-\alpha \log \alpha \geq -C(\gamma \alpha) \log(\gamma \alpha).$$

Indeed, for $\gamma < 1$ fixed we have

(65) $$g(\alpha) = \frac{(\gamma \alpha) \log(\gamma \alpha)}{\alpha \log \alpha} = \gamma + \frac{\gamma \log \gamma}{\log \alpha}$$

is increasing for $\alpha \in [0, 1/2]$. Hence it suffices to choose a $\gamma > 0$ sufficiently small so that

$$g(1/2) = \frac{(1/2) \log(1/2)}{(\gamma/2) \log(\gamma/2)} \geq C,$$

so that

$$g(\alpha) \geq g(1/2) \geq C,$$

which along with equation (65) yields equation (64).

It follows that for sufficiently small $\gamma > 0$ we have

$$\frac{\log(bp^2)}{n} \geq -(\alpha/2) \log(\alpha/2) + O\left(|\delta \log \delta| + \frac{\log n}{n}\right),$$

and, since $\delta \leq \gamma a/n = \gamma \alpha$, this expression is

$$\geq -(\alpha/4) \log(\alpha/2) + O\left(\frac{\log n}{n}\right).$$

Hence for any constant, C_1, there is a C_2 such that if $a \geq C_2$ (i.e., $\alpha \geq C_2/n$) then

$$\frac{\log(bp^2)}{n} \leq \frac{-C_1 \log n}{n}$$

for all n sufficiently large. In other words $bp^{d/2}$, i.e., the expression in equation (58), is at most n^{-C_1}; this, by the discussion after equation (58), completes the proof. □

THEOREM 12.4. *Theorem 12.3 holds in the models $\mathcal{H}_{n,d}$, $\mathcal{I}_{n,d}$, and $\mathcal{J}_{n,d}$.*

PROOF. In $\mathcal{H}_{n,d}$ each permutation occurs with probability at most n times its probability in $\mathcal{G}_{n,d}$. Therefore the same analysis goes through, except that p is multiplied by at most a factor of n. This changes the expression for $n^{-1} \log(bp^2)$ by an $O(n^{-1} \log n)$ factor, so the same proof carries over.

For $\mathcal{I}_{n,d}$ we again set C_i and c_i as before. A perfect matching in $\{1, \ldots, n\}$ will have (1) r vertices of C_1 paired amongst themselves, (2) $c_1 - r$ vertices of C_1 paired with C_3 vertices, and (3) c_2 vertices of C_2 paired with C_3 vertices. This data determines the pairing for $r + 2(c_1 - r) + 2c_2$ vertices. The expression for b, representing the number of ways the C_i can be chosen, is the same as before. We now derive an expression for p, the probability that a single perfect matching matches all A vertices to those in B.

12. MAGNIFICATION THEOREMS

For an even integer, m, let m *odd factorial* be
$$m!_{\text{odd}} = (m-1)(m-3)\cdots 3 = \frac{m!}{2^{m/2}(m/2)!},$$
which is just the number of perfect matchings of m elements. Stirling's formula yields
$$m!_{\text{odd}} \sim \sqrt{2}\,(m/e)^{m/2}$$
(so that for our purposes $m!_{\text{odd}}$ can be regarded as replaceable by the square root of $m!$).

We have
$$p = p(\{c_i\}, r, n) = \left[\binom{c_1}{r} r!_{\text{odd}}\right] \left[\binom{c_3}{c_1 - r}(c_1 - r)!\right]$$
$$\times \left[\binom{c_3 - c_1 + r}{c_2} c_2!\right] \frac{(n - 2c_1 - 2c_2 - r)!_{\text{odd}}}{n!_{\text{odd}}}.$$

We get that $-\log p$ is
$$\frac{h(\nu_1, \nu_2)}{2} + O\left(|\delta \log \delta| + \frac{\log n}{n}\right),$$
with h as in equation (63). Since b is unchanged, by analyzing as before we see that there is a fixed $\gamma > 0$ such that for any constant C_1 there is a C_2 such that $bp^3 = O(n^{-C_1})$ provided that $a \geq C_2$.

Next we consider $\mathcal{J}_{n,d}$. Let G be a random graph in $\mathcal{J}_{n,d}$, so $V_G = \{1, \ldots, n\}$. Consider the graph G' formed by adding one new vertex, $w = n+1$, to G and replacing each half-loop about a vertex, v, in G by an edge from v to w. Then G' is precisely distributed as an element of $\mathcal{I}_{n+1,d}$; indeed, a perfect matching on $V_{G'}$ matches w to some element of $V_G = \{1, \ldots, n\}$ and then randomly matches the remaining $n-1$ elements of V_G.

Now we know that G' is a γ-spreader with probability $1 - O(n^{-\tau_{\text{fund}}})$. But for any $A \subset V_G$, $\Gamma_{G'}(A)$ consists of at most one more vertex than $\Gamma_G(A)$. Hence for $|A| \leq |V_G|/2$ and G' being a γ-spreader, we have
$$|\Gamma_G(A) \setminus A| \geq \gamma |A| - 1 \geq \gamma' |A|,$$
where $\gamma' = \gamma - (1/c')$, provided that $|A| \geq c'$. Hence G is a γ-spreader on sets, A, of size $\max(c, c') \leq |A| \leq n/2$. On smaller sets, A, we have G is a γ'-spreader with probability $1 - O(n^{-\tau_{\text{fund}}})$, assuming $\gamma' \leq 1/\max(c, c')$, by the argument given before for $\mathcal{G}_{n,d}$. (Notice that the τ_{fund} for $\mathcal{I}_{n,d}$ and $\mathcal{J}_{n,d}$ are the same.) Hence a random graph in $\mathcal{J}_{n,d}$ is a γ'-spreader with probability $1 - O(n^{-\tau_{\text{fund}}})$. □

CHAPTER 13

Finishing the $\mathcal{G}_{n,d}$ Proof

Here we quickly finish the proof of Theorem 1.1, which proves Alon's conjecture for $\mathcal{G}_{n,d}$.

Fix a value of r to be specified later. Let $\Psi' = \Psi_{\text{eig}}[r-1]$ be the set of supercritical tangles of order less than r, and let $\Psi = \Psi_{\min}[r-1]$ (which we recall is the set of minimal $\Psi' = \Psi_{\text{eig}}[r-1]$ elements with respect to inclusion); we know that Ψ is finite by Lemma 9.2, and we recall that if G contains a Ψ' tangle then it contains an element of Ψ. The probability that $\chi_\Psi(G) = 1$, i.e., that G contains at least one element of Ψ, is at most $O(n^{-\tau_{\text{fund}}})$, since every one of the finitely many tangles in Ψ occurs with probability proportional to $1/n$ to the order of tangle (Theorem 4.7), and each tangle order is at least τ_{fund}. Given that $\chi_\Psi(G) = 0$, we have that G contains no supercritical tangle of order less than r, and hence no irreducible closed walk can fail to be (S, Ψ')-selective for any S. Hence for all S and k we have

$$\chi_\Psi(G) = 0 \quad \text{implies} \quad \text{SSIT}_{S,\Psi'}(G;k) = \text{SIT}(G;k).$$

Thus

$$\mathbf{E}\left[(1-\chi_\Psi)\text{SSIT}_{S,\Psi'}(G;k)\right] = \mathbf{E}\left[(1-\chi_\Psi)\text{SIT}(G;k)\right].$$

Now according to Theorem 10.3 we have

$$\text{SIT}(G;k) = \sum_{i=1}^{n} \mu_1^k(\lambda_i) + \mu_2^k(\lambda_i) + \left(1+(-1)^k\right)(d-2)/2,$$

for even $k \geq 2$, where

$$\mu_{1,2}(\lambda) = \frac{\lambda \pm \sqrt{\lambda^2 - 4(d-1)}}{2}.$$

In other words, there are nd numbers ν_i, such that $\text{SIT}(G;k)$ is the sum of the k-th powers of these numbers. Also for each i we have ν_i is not real only if it is of absolute value $\sqrt{d-1}$. Combining this and Theorem 9.3 we see that

$$\theta_i = 1 - (1-\chi_\Psi)\nu_i/(d-1)$$

are random variables that satisfy the conditions of Lemma 11.1 for each i (and G). It follows that
(66)
$$\mathbf{E}\left[(1-\chi_\Psi)\sum_{i=1}^{n}\sum_{\substack{j \text{ such that}\\|\mu_j(\lambda_i)|\leq(d-1)(1-\log^{-2}n)}} \mu_j(\lambda_i)^k\right] = O(Dn^{1-(r/3)} + k^c(d-1)^{-k/2})$$

for all k with $1 \leq k \leq n^\gamma$ for some constant $\gamma > 0$ depending only on r.

According to Theorems 12.2 and 12.3 there is an $\epsilon > 0$ such that with probability $1 - O(n^{-\tau_{\text{fund}}})$ we have $|\lambda_i| \le d - \epsilon$ for all $i \ne 1$; in this case there is an $\epsilon' = \epsilon'(\epsilon) > 0$ such that $|\mu_j(\lambda_i)| \le (d-1) - \epsilon'$ for all $j = 1, 2$ and $i \ne 1$.

We claim that for any G and an even integer k we have
$$\sum_{i \text{ s.t. } \mu_{1,2}(\lambda_i) \text{ not real}} \sum_{j=1}^{2} \mu_j(\lambda_i)^k \ge -2(n-1)(d-1)^{k/2}$$
indeed, if $\mu_j(\lambda_i)$ is not real, it is of absolute value $\sqrt{d-1}$; if $\mu_j(\lambda_i)$ is real then its k-th power is non-negative.

Now let A be the event that $\chi_\Psi = 0$ and that $|\mu_j(\lambda_i)| \le (d-1) - \epsilon'$ for all $j = 1, 2$ and $i \ne 1$. Let $B = B(\eta)$ be the event that for some j and some $i \ne 1$ we have $|\mu_j(\lambda_i)| \ge e^\eta \sqrt{d-1}$ for an arbitrary fixed $\eta > 0$. $A \cap B$ implies that for even integer k we have
$$\sum_{i=2}^{n} \sum_{j=1}^{2} \mu_j(\lambda_i)^k \ge \left(e^\eta \sqrt{d-1}\right)^k - 2(n-2)(d-1)^{k/2}.$$

It follows, using equation (66), that for even k,
$$\text{Prob}\{A \cap B\} \left(e^\eta \sqrt{d-1}\right)^k \le \mathrm{E}\left[\sum_{i \text{ s.t. } \mu_{1,2}(\lambda_i) \text{ real}} \sum_{j=1}^{2} \mu_j(\lambda_i)^k\right]$$
$$= \mathrm{E}\left[\sum_{i=2}^{n} \sum_{j=1}^{2} \mu_j(\lambda_i)^k\right] - \mathrm{E}\left[\sum_{i \text{ s.t. } \mu_{1,2}(\lambda_i) \text{ not real}} \sum_{j=1}^{2} \mu_j(\lambda_i)^k\right]$$
$$\le O(Dn^{1-(r/3)}(d-1)^k + k^c(d-1)^{k/2}) + 2(n-1)(d-1)^{k/2}.$$

We now take
$$k = 2\left\lceil \frac{r \log n}{3 \log(d-1)} \right\rceil.$$

We have
$$(k/2) - 1 \le \frac{r \log n}{3 \log(d-1)} \le k/2.$$

Hence
$$n^{-r/3} \le (d-1)^{-(k/2)+1},$$
and so
$$\text{Prob}\{A \cap B\} \le c \max(k^c, n) e^{-k\eta}$$
$$\le cne^{-k\eta} = cnn^{-\alpha r},$$
where $\alpha = (2/3)\eta/\log(d-1)$, i.e. α is a positive constant (depending only on η and d). Choosing r so that $\alpha r - 1 > \tau_{\text{fund}}$, we have
$$\text{Prob}\{A \cap B\} = O(n^{-\tau_{\text{fund}}}).$$

But we have already seen (Theorems 12.2 and 12.3) that
$$\text{Prob}\{A^c\} = O(n^{-\tau_{\text{fund}}}),$$
where A^c is the complement of A. Hence
$$\text{Prob}\{B\} = \text{Prob}\{B \cap A\} + \text{Prob}\{B \cap A^c\} = O(n^{-\tau_{\text{fund}}}).$$

13. FINISHING THE $\mathcal{G}_{n,d}$ PROOF

For any $\epsilon > 0$ there is an $\eta > 0$ such that $|\lambda| \geq 2\sqrt{d-1} + \epsilon$ implies $|\mu_i(\lambda)| \geq e^\eta \sqrt{d-1}$ for at least one i, which is the event $B = B(\eta)$ above. It follows that for any $\epsilon > 0$ we have

$$\text{Prob}\left\{|\lambda_i| \geq 2\sqrt{d-1} + \epsilon \text{ for some } i > 1\right\} = O(n^{-\tau_{\text{fund}}}).$$

This (and Theorem 2.11) proves Theorem 1.1.

CHAPTER 14

Finishing the Proofs of the Main Theorems

We now complete the proofs of Theorems 1.2 and 1.3, i.e., we establish the Alon conjecture for $\mathcal{H}_{n,d}$, $\mathcal{I}_{n,d}$, and $\mathcal{J}_{n,d}$.

The proofs of the theorems are as the proof for $\mathcal{G}_{n,d}$. We only need to establish the following results for the different models of random graph:

(1) Labelling: The model comes with edges labelled from a set Π such that to each $\pi \in \Pi$ we associate a $\pi^{-1} \in \Pi$ such that $(\pi^{-1})^{-1} = \pi$ (in other words, the elements of Π are paired, with the possibility that an element is paired with itself).

(2) Coincidence: If k of the random edges have been determined, and if we fix any two vertices, v, w, in the graph, then the probability that an edge of a given label takes v to w is at most $c/(n - ck)$ for some constant c. We have only briefly mentioned coincidences in this paper, but our Lemmas 5.7 and 5.8, proven in [**Fri91**], require a property like this.

(3) Expansion with Error: Consider a Π-labelled graph, H, with vertices a subset of $\{1, \ldots, n\}$, that can occur as a subgraph of a graph in our model. The probability that H occurs must depend only on the number of edges, a_π, of each label, π (of course, $a_\pi = a_{\pi^{-1}}$). Furthermore, this probability times the number subsets of $\{1, \ldots, n\}$ of size V_H is, for every positive integer r,

$$\mathrm{E}_{\mathrm{symm}}(H)_n = \left(\sum_{i=0}^{r-1} \frac{p_i(\vec{a})}{n^i}\right) + \frac{\mathrm{error}}{n^r},$$

where p_i are polynomials in $\vec{a}$ (where $\vec{a}$ is the collection of all a_π) and where

$$|\mathrm{error}| \le ck^{r'}$$

for all $k \le n/c$, where c_1, r' depend only on r. Furthermore, $p_i = 0$ if i is less than the order of H.

(4) Simple Word Sum: Let $\mathrm{Irred}_{k,\sigma,\tau}$ be those words that begin with σ, end in τ, and are *irreducible* (meaning no consecutive occurrence of π and π^{-1}). Then for any polynomial, $p = p(\vec{a})$ (with $\vec{a}$ as above), we require

(67) $$\sum_{w \in \mathrm{Irred}_{k,\sigma,\tau}} p\bigl(a_1(w), \ldots, a_{d/2}(w), k\bigr) = (d-1)^k Q_1(k) + E(k)$$

for a polynomial, Q_1, and a function E with $|E(k)| \le ck^c$ for some constant c (i.e., the above sum is super-d-Ramanujan).

(5) τ_{fund} determination: We must determine τ_{fund} for the model.

(6) Spreading: There is a constant $\gamma > 0$ such that the probability that a random graph has $|\lambda_i| \ge d - \gamma$ for some $i \ne 1$ is of order at most $n^{-\tau_{\mathrm{fund}}}$.

We have already shown spreading and determined τ_{fund} for all three models. The labelling of the models is quite simple: $\mathcal{H}_{n,d}$ is labelled like $\mathcal{G}_{n,d}$; $\mathcal{I}_{n,d}$ is labelled with $\Sigma = \{\sigma_1, \ldots, \sigma_d\}$ where $\sigma_i^{-1} = \sigma_i$ (each σ_i represents a perfect matching); $\mathcal{J}_{n,d}$ is labelled with $\Sigma \cap T$ with Σ as before and $T = \{\tau_1, \ldots, \tau_d\}$ with $\tau_i^{-1} = \tau_i$, and where the σ_i represent the near perfect matching and the τ_i represents the single completing half-loop for σ_i.

Coincidence is easily checked for all three models.

We address the issue of Simple Word Sum. The word sum for $\mathcal{H}_{n,d}$ is the same as for $\mathcal{G}_{n,d}$. For $\mathcal{I}_{n,d}$, the technique of Lemma 2.11 of [**Fri91**] reduces the matter to the irreducible eigenvalues of a vertex with d half-loops; since these eigenvalues are the eigenvalues of a $d \times d$ matrix which is 0 on the diagonal and 1's elsewhere, the eigenvalues are $d-1$ with multiplicity 1 and -1 with multiplicity $d-1$. Hence the simple word sum of equation (67) is given by

$$(68) \qquad (d-1)^k Q_1(k) + (-1)^k Q_2(k)$$

where Q_i are polynomials. For $\mathcal{J}_{n,d}$ we can break the sum by how many half-loops are involved. For a fixed set of half-loops involved in the irreducible word, the sum is a convolution of functions of the form in equation (68), which by Theorem 7.2 is again super-d-Ramanujan.

We now establish Expansion with Error for the three models. Equation (16) has the $\mathcal{H}_{n,d}$ analogue

$$P(w; \vec{t}) = \prod_{i=1}^{d/2} \frac{(n - a_i - 1)!}{(n-1)!}.$$

Now recall the proof of Theorem 5.5, especially equations (19) and (20). For $\mathcal{H}_{n,d}$, we have

$$(69) \qquad \mathrm{E}_{\text{symm}}(H)_n = n(n-1)\cdots(n-v+1) \prod_{i=1}^{d/2} \frac{(n-a_i-1)!}{(n-1)!}.$$

$$= n^{v-e} g(1/n),$$

with g as in equation (19) with $b_1, \ldots, b_v$ being $0, 1, \ldots, v-1$ and $c_1, \ldots, c_e$ being the collection of the sequences $1, 2, \ldots, a_i$. Hence for a walk of length at most k we have

$$\sum b_j + \sum c_j \le \binom{k}{2} + \binom{k+1}{2} = k^2.$$

Accordingly Expansion with Error holds for $\mathcal{H}_{n,d}$ with expansion polynomials determined by equation (69), and with error term bounded by

$$e^{rk/(n-k)} k^{2r};$$

this bound is $\le ck^{r'}$ for all $k \le n$ with $r' = 2r$ and $c = e^r$.

Similarly for $\mathcal{I}_{n,d}$ we have the analogue

$$P(w; \vec{t}) = \prod_{i=1}^{d} \frac{(n-a_i)!_{\text{odd}}}{n!_{\text{odd}}}.$$

The analysis goes through essentially as before; in the error bound we have $\sum b_j$ is again $\binom{k}{2}$, but this time the $\sum c_j$ is as large as

$$1 + 3 + 5 + \cdots + (2k-1) = k^2$$

14. FINISHING THE PROOFS OF THE MAIN THEOREMS

(taking one $a_i = 2k$ and the rest 0). So Expansion with Error holds for $\mathcal{I}_{n,d}$ with error term bounded by

$$e^{rk/n}\left(k^2 + \binom{k}{2}\right)^r \le e^{rk/n}(2k)^{2r}.$$

For $\mathcal{J}_{n,d}$, consider a random 1-regular graph, G', consisting of a near perfect matching plus one complementing half-loop on the vertex set $\{1,\ldots,n\}$. Notice that the number of such graphs is $n(n-1)!_{\text{odd}}$. Hence the probability of occurrence of a specified half-loop and a other matchings in G' is

$$\frac{(n-1-2a)!_{\text{odd}}}{n(n-1)!_{\text{odd}}} = \frac{1}{n(n-2)\cdots(n-2a)},$$

and the probability of a specified matchings (with no specified half-loop) is

$$\frac{(n-2a)(n-1-2a)!_{\text{odd}}}{n(n-1)!_{\text{odd}}} = \frac{1}{n(n-2)\cdots(n-2a-2)}.$$

So for any specification of half-loops in H, i.e., any fixing of each a_{τ_i} to 0 or 1, $\mathrm{E}_{\text{symm}}(H)_n$ is a polynomial in the a_{σ_i}'s; this makes $\mathrm{E}_{\text{symm}}(H)_n$ a polynomial in the $\vec{a}$, namely

$$\sum_{I \subset \{1,\ldots,d\}} \left(p_I(a_{\sigma_1},\ldots,a_{\sigma_d}) \prod_{i \in I} a_{\tau_i} \prod_{i \notin I}(1-a_{\tau_i}) \right).$$

We also see that, in the terminology above, $\sum b_j = \binom{k}{2}$ and $\sum c_j \le 2\binom{k-1}{2}$. Hence Expansion with Error holds for $\mathcal{J}_{n,d}$ as well.

This establishes the six required results mentioned at the beginning of this section for the models $\mathcal{H}_{n,d}$, $\mathcal{I}_{n,d}$, and $\mathcal{J}_{n,d}$. Theorems 1.2 and 1.3 follow.

CHAPTER 15

Closing Remarks

We make a number of final remarks.

Stronger conjectures: As mentioned before, numerical experiments indicate that the average (and median) λ_2 for a random graph is $2\sqrt{d-1}+\epsilon(n)$, where $\epsilon(n)$ is a negative function (tending to 0 as $n \to \infty$). By the results of Friedman and Kahale (extending the Alon-Boppana result), $-\epsilon(n) \le O(\log^{-2} n)$ (see [**Fri93**]). However, the trace method, even with selective traces, seems to require some fundamental new idea in order to have any hope of achieving $\epsilon(n)$ that is zero or negative.

Critical d: As mentioned before, when there is a critical tangle of order strictly less than that of any hypercritical tangle, then our techniques leave a gap in that we can only prove $\lambda_2 > 2\sqrt{d-1}$ with probability at least c/n^s where $s > \tau_{\text{fund}}$. This case is extremely interesting, since it seems that there should be a theorem that closes this gap, and such a theorem would either get around a poorly bounded $W_{\widetilde{T},S}(M_1, M_2)$ or improve the very interesting Theorem 3.13 (or do something else).

Relative Alon Conjecture: Following [**Fri03**], it seems quite possible to relativize the main theorems in this paper. Namely, fix a "base" graph, B, (or, more generally, a "base" pregraph, in the sense of [**Fri93**]). Fix an $\epsilon > 0$. Then we believe that most random coverings of B of degree n have all "new" eigenvalue $\le \epsilon + \rho$, where ρ is the spectral radius of the universal cover of B. Similarly, we can ask for ϵ to be zero or even a negative function of n. See [**Fri03, FT02**] for further discussion and a result in this direction.

Alternate Proof with Trace (see the end of Chapter 2): It may be possible to analyze the expected irreducible trace over all of $\mathcal{G}_{n,d}$. As remarked in Theorem 2.12 and the discussion thereafter, the coefficients of $g_i(k)$ there could no longer be d-Ramanujan. It may be possible to analyze selective traces without discarding contributions from tangled graphs. In other words, if we better understood how selectivity affected irreducible traces, we might not need Chapter 9 (and certain parts of our understanding of these traces might improve). Clearly selectivity in G can be expressed in terms of walks in an induced subgraph of a "higher block presentation" of G (see [**LM95, Kit98**]). However, it is not clear what can be said about the eigenvalues of induced subgraphs of a higher block presentation; the author has only some weak results in this directions (see [**Fri**]).

Glossary

This glossary contains a term or a piece of notation, followed by a colon (:), followed by a brief description, followed by the page number(s) where the term/notation is explained.

$\lambda_i(G)$: i-th largest eigenvalue of the adjacency matrix of a finite graph, page 3

$\mathcal{G}_{n,d}$: space of random graphs formed from $d/2$ permutations, page 3

τ_{fund}: smallest order of a supercritical tangle, page 3

τ_{fund}: smallest order of a supercritical tangle, page 4

$\mathcal{H}_{n,d}$: random graph space formed by $d/2$ permutations that are cycles of length n, page 4

$\mathcal{I}_{n,d}$: random graph model of d perfect matchings (n even), page 4

$\mathcal{J}_{n,d}$: random graph model formed from d permutations each with exactly one fixed point (n odd), page 4

closed walk: a walk beginning and ending at the same vertex, page 5

irreducible: a walk (resp. word) that has no consecutive steps of an edge (resp. letter) and its inverse, page 5

IrredTr (A, k): the number of closed irreducible walks of length k in the graph underlying A, page 10

graph: a directed graph with a origin/terminal reversing pairing of its edges, page 19

half-loop: a self-loop paired with itself (in a graph), page 19

whole-loop: two self-loops (about the same vertex) paired with each other (in a graph), page 19

variable-length graph (VLG): a graph, directed or undirected, with a positive integral "length" associated to each edge, page 20

VLG: variable-length graph, page 20

bead: a vertex with indegree and outdegree 1 (or, for undirected graphs, degree 2) with no self-loops, page 20

$c_G(u, v; k)$: the number of walks of length k from u to v in G, page 21

$\lambda_1(G)$: the sup over all values $(c_G(v, v; k))^k$, page 21

$\lambda_1(G)$: the sup over all values $(c_G(v, v; k))^k$, page 21

generating function: a power series, $\Sigma_k a_k z^k$, formed from coefficients a_k, page 22

G_{Irred}: edge graph of G with edges joined only when they form an irreducible path, page 24

$\lambda_{\text{Irred}}(G)$: the largest eigenvalue of G_{Irred}, page 24

critical: a tangle with $\lambda_{\text{Irred}} = \sqrt{d-1}$, page 30

supercritical: a tangle with $\lambda_{\text{Irred}} \geq \sqrt{d-1}$, page 30

hypercritical: a tangle with $\lambda_{\text{Irred}} > \sqrt{d-1}$, page 30

a_j: the number of π_j and π_j^{-1} appearing in a word or form, page 38

expansion polynomials: The polynomials $p_i = p_i(a_1,\ldots,a_{d/2},v)$ giving the $1/n$ expansion of $\mathrm{E}_{\mathrm{symm}}(w,\vec{t})$, page 38

$p_i = p_i(a_1,\ldots,a_{d/2},v)$: the expansion polynomials, page 38

a_j: the number of π_j and π_j^{-1} appearing in a word or form, page 38

a_j: the number of π_j and π_j^{-1} appearing in a word or form, page 41

form: an oriented, Π-labelled graph with edges and vertices numbered that is meant to represent the graph traced out by a potential word, forgetting about the particular choices in $\{1,\ldots,n\}$ of the vertices, page 42

$\mathrm{E}[\Gamma]_n$: The expected value of the number of closed walks corresponding to a potential walk class (which depends only the the form of the potential walk), page 43

a_j: the number of π_j and π_j^{-1} appearing in a word or form, page 43

$W_\Gamma, W_T, W_{T'}$: the number of potential walk classes or legal walks on a form, type, or new type with various additional restrictions (such as irreducibility), page 43

type: a graph representing a number of forms, where we forget certain features of the form, such as its Π-labelling and all or almost all its degree two vertices, page 44

new type: a type with some additional information specified, such as a partition of the types edges into a "long" and a "fixed" edge set, page 45

$\mathrm{IrSelTr}_{S,\Psi}(G;k)$: k-th irreducible (S,Ψ)-selective trace of G, i.e., the number of irreducible closed walks of length k in G such that no subpath of length at most S traces out a tangle in Ψ, page 49

$W_\Gamma, W_T, W_{T'}$: the number of potential walk classes or legal walks on a form, type, or new type with various additional restrictions (such as irreducibility), page 50

Ω-type: a generalization of type that incorporates information of an included tangle, Ω, page 71

Bibliography

[AFKM86] R. Adler, J. Friedman, B. Kitchens, and B. Marcus. State splitting for variable length graphs. *IEEE Transactions on Information Theory*, 32(1):108–115, January 1986.

[Alo86] N. Alon. Eigenvalues and expanders. *Combinatorica*, 6(2):83–96, 1986. Theory of computing (Singer Island, Fla., 1984).

[BS87] Andrei Broder and Eli Shamir. On the second eigenvalue of random regular graphs. In *28th Annual Symposium on Foundations of Computer Science*, pages 286–294, 1987.

[Buc86] Marshall W. Buck. Expanders and diffusers. *SIAM J. Algebraic Discrete Methods*, 7(2):282–304, 1986.

[DD89] Warren Dicks and M. J. Dunwoody. *Groups acting on graphs*. Cambridge University Press, Cambridge, 1989.

[Dod84] Jozef Dodziuk. Difference equations, isoperimetric inequality and transience of certain random walks. *Trans. Amer. Math. Soc.*, 284(2):787–794, 1984.

[FK81] Z. Füredi and J. Komlós. The eigenvalues of random symmetric matrices. *Combinatorica*, 1(3):233–241, 1981.

[FKS89] J. Friedman, J. Kahn, and E. Szemerédi. On the second eigenvalue of random regular graphs. In *21st Annual ACM Symposium on Theory of Computing*, pages 587–598, 1989.

[Fri] Joel Friedman. SVD misalignment and eigenvalue separation. In preparation.

[Fri91] Joel Friedman. On the second eigenvalue and random walks in random d-regular graphs. *Combinatorica*, 11(4):331–362, 1991.

[Fri93] Joel Friedman. Some geometric aspects of graphs and their eigenfunctions. *Duke Math. J.*, 69(3):487–525, 1993.

[Fri03] Joel Friedman. Relative expanders or weakly relatively Ramanujan graphs. *Duke Math. J.*, 118(1):19–35, 2003.

[FT02] Joel Friedman and Jean-Pierre Tillich. Generalized Alon-Boppana theorems and error-correcting codes. May 2002. preprint.

[FTP83] Alessandro Figà-Talamanca and Massimo A. Picardello. *Harmonic analysis on free groups*. Marcel Dekker Inc., New York, 1983.

[Gem80] S. Geman. A limit theorem for the norm of random matrices. *Ann. of Prob.*, 8(2):252–261, 1980.

[GJKW02] Catherine Greenhill, Svante Janson, Jeong Han Kim, and Nicholas C. Wormald. Permutation pseudographs and contiguity. *Combin. Probab. Comput.*, 11(3):273–298, 2002.

[God93] C. D. Godsil. *Algebraic combinatorics*. Chapman & Hall, New York, 1993.

[HMS91] Chris D. Heegard, Brian H. Marcus, and Paul H. Siegel. Variable-length state splitting with applications to average runlength-constrained (ARC) codes. *IEEE Trans. Inform. Theory*, 37(3, part 2):759–777, 1991.

[Joh90] D. L. Johnson. *Presentations of groups*. Cambridge University Press, Cambridge, 1990.

[JS89] Mark Jerrum and Alistair Sinclair. Approximating the permanent. *SIAM J. Comput.*, 18(6):1149–1178, 1989.

[Kit98] Bruce P. Kitchens. *Symbolic dynamics*. Springer-Verlag, Berlin, 1998.

[KW01] Jeong Han Kim and Nicholas C. Wormald. Random matchings which induce Hamilton cycles and Hamiltonian decompositions of random regular graphs. *J. Combin. Theory Ser. B*, 81(1):20–44, 2001.

[LM95] Douglas Lind and Brian Marcus. *An introduction to symbolic dynamics and coding*. Cambridge University Press, Cambridge, 1995.

[LPS86] A. Lubotzky, R. Phillips, and P. Sarnak. Explicit expanders and the ramanujan conjectures. In *18th Annual ACM Symposium on Theory of Computing*, pages 240–246, 1986.

[LPS88] A. Lubotzky, R. Phillips, and P. Sarnak. Ramanujan graphs. *Combinatorica*, 8(3):261–277, 1988.

[Mar88] G. A. Margulis. Explicit group-theoretic constructions of combinatorial schemes and their applications in the construction of expanders and concentrators. *Problemy Peredachi Informatsii*, 24(1):51–60, 1988.

[McK81] B. McKay. The expected eigenvalue distribution of a large regular graph. *Lin. Alg. Appl.*, 40:203–216, 1981.

[Mor94] Moshe Morgenstern. Existence and explicit constructions of $q+1$ regular Ramanujan graphs for every prime power q. *J. Combin. Theory Ser. B*, 62(1):44–62, 1994.

[Nil91] A. Nilli. On the second eigenvalue of a graph. *Discrete Math.*, 91(2):207–210, 1991.

[Sen81] E. Seneta. *Nonnegative matrices and Markov chains*. Springer Series in Statistics. Springer-Verlag, New York, second edition, 1981.

[SJ89] Alistair Sinclair and Mark Jerrum. Approximate counting, uniform generation and rapidly mixing Markov chains. *Inform. and Comput.*, 82(1):93–133, 1989.

[SW49] Claude E. Shannon and Warren Weaver. *The Mathematical Theory of Communication*. University of Illinois Press, 1971/1949.

[Wig55] E. Wigner. Characteristic vectors of bordered matrices with infinite dimensions. *Annals of Math.*, 63(3):548–564, 1955.

[Woe00] Wolfgang Woess. *Random walks on infinite graphs and groups*. Cambridge University Press, Cambridge, 2000.

[Wor99] N. C. Wormald. Models of random regular graphs. In *Surveys in combinatorics, 1999 (Canterbury)*, pages 239–298. Cambridge Univ. Press, Cambridge, 1999.

Editorial Information

To be published in the *Memoirs*, a paper must be correct, new, nontrivial, and significant. Further, it must be well written and of interest to a substantial number of mathematicians. Piecemeal results, such as an inconclusive step toward an unproved major theorem or a minor variation on a known result, are in general not acceptable for publication.

Papers appearing in *Memoirs* are generally at least 80 and not more than 200 published pages in length. Papers less than 80 or more than 200 published pages require the approval of the Managing Editor of the Transactions/Memoirs Editorial Board.

As of May 31, 2008, the backlog for this journal was approximately 17 volumes. This estimate is the result of dividing the number of manuscripts for this journal in the Providence office that have not yet gone to the printer on the above date by the average number of monographs per volume over the previous twelve months, reduced by the number of volumes published in four months (the time necessary for preparing a volume for the printer). (There are 6 volumes per year, each usually containing at least 4 numbers.)

A Consent to Publish and Copyright Agreement is required before a paper will be published in the *Memoirs*. After a paper is accepted for publication, the Providence office will send a Consent to Publish and Copyright Agreement to all authors of the paper. By submitting a paper to the *Memoirs*, authors certify that the results have not been submitted to nor are they under consideration for publication by another journal, conference proceedings, or similar publication.

Information for Authors

Memoirs are printed from camera copy fully prepared by the author. This means that the finished book will look exactly like the copy submitted.

Initial submission. The AMS uses Centralized Manuscript Processing for initial submissions. Authors should submit a PDF file using the Initial Manuscript Submission form found at www.ams.org/cgi-bin/peertrack/submission.pl, or send one copy of the manuscript to the following address: Centralized Manuscript Processing, MEMOIRS OF THE AMS, 201 Charles Street, Providence, RI 02904-2294 USA. If a paper copy is being forwarded to the AMS, indicate that it is for it Memoirs and include the name of the corresponding author, contact information such as email address or mailing address, and the name of an appropriate Editor to review the paper (see the list of Editors below).

The paper must contain a *descriptive title* and an *abstract* that summarizes the article in language suitable for workers in the general field (algebra, analysis, etc.). The *descriptive title* should be short, but informative; useless or vague phrases such as "some remarks about" or "concerning" should be avoided. The *abstract* should be at least one complete sentence, and at most 300 words. Included with the footnotes to the paper should be the 2000 *Mathematics Subject Classification* representing the primary and secondary subjects of the article. The classifications are accessible from www.ams.org/msc/. The list of classifications is also available in print starting with the 1999 annual index of *Mathematical Reviews*. The Mathematics Subject Classification footnote may be followed by a list of *key words and phrases* describing the subject matter of the article and taken from it. Journal abbreviations used in bibliographies are listed in the latest *Mathematical Reviews* annual index. The series abbreviations are also accessible from www.ams.org/publications/. To help in preparing and verifying references, the AMS offers MR Lookup, a Reference Tool for Linking, at www.ams.org/mrlookup/.

Electronically prepared manuscripts. The AMS encourages electronically prepared manuscripts, with a strong preference for $\mathcal{AMS}$-LATEX. To this end, the Society has prepared $\mathcal{AMS}$-LATEX author packages for each AMS publication. Author packages include instructions for preparing electronic manuscripts, samples, and a style file that generates the particular design specifications of that publication series. Though $\mathcal{AMS}$-LaTeX is the highly preferred format of TeX, author packages are also available in $\mathcal{AMS}$-TeX.

Authors may retrieve an author package from the AMS website starting from www.ams.org/tex/ or via FTP to ftp.ams.org (login as anonymous, enter username as password, and type cd pub/author-info). The *AMS Author Handbook* and the *Instruction Manual* are available in PDF format following the author packages link from www.ams.org/tex/. The author package can also be obtained free of charge by sending email to tech-support@ams.org (Internet) or from the Publication Division, American Mathematical Society, 201 Charles St., Providence, RI 02904-2294, USA. When requesting an author package, please specify $\mathcal{AMS}$-LaTeX or $\mathcal{AMS}$-TeX and the publication in which your paper will appear. Please be sure to include your complete mailing address.

After acceptance. The final version of the electronic file should be sent to the Providence office (this includes any TeX source file, any graphics files, and the DVI or PostScript file) immediately after the paper has been accepted for publication.

Before sending the source file, be sure you have proofread your paper carefully. The files you send must be the EXACT files used to generate the proof copy that was accepted for publication. For all publications, authors are required to send a printed copy of their paper, which exactly matches the copy approved for publication, along with any graphics that will appear in the paper.

Accepted electronically prepared files can be submitted via the web at www.ams.org/submit-book-journal/, sent via FTP, or sent on CD-Rom or diskette to the Electronic Prepress Department, American Mathematical Society, 201 Charles Street, Providence, RI 02904-2294 USA. TeX source files, DVI files, and PostScript files can be transferred over the Internet by FTP to the Internet node ftp.ams.org (130.44.1.100). When sending a manuscript electronically via CD-Rom or diskette, please be sure to include a message identifying the paper as a Memoir.

Electronically prepared manuscripts can also be sent via email to pub-submit@ams.org (Internet). In order to send files via email, they must be encoded properly. (DVI files are binary and PostScript files tend to be very large.)

Electronic graphics. Comprehensive instructions on preparing graphics are available at www.ams.org/jourhtml/. A few of the major requirements are given here.

Submit files for graphics as EPS (Encapsulated PostScript) files. This includes graphics originated via a graphics application as well as scanned photographs or other computer-generated images. If this is not possible, TIFF files are acceptable as long as they can be opened in Adobe Photoshop or Illustrator. No matter what method was used to produce the graphic, it is necessary to provide a paper copy to the AMS.

Authors using graphics packages for the creation of electronic art should also avoid the use of any lines thinner than 0.5 points in width. Many graphics packages allow the user to specify a "hairline" for a very thin line. Hairlines often look acceptable when proofed on a typical laser printer. However, when produced on a high-resolution laser imagesetter, hairlines become nearly invisible and will be lost entirely in the final printing process.

Screens should be set to values between 15% and 85%. Screens which fall outside of this range are too light or too dark to print correctly. Variations of screens within a graphic should be no less than 10%.

Inquiries. Any inquiries concerning a paper that has been accepted for publication should be sent to memo-query@ams.org or directly to the Electronic Prepress Department, American Mathematical Society, 201 Charles St., Providence, RI 02904-2294 USA.

Editors

This journal is designed particularly for long research papers, normally at least 80 pages in length, and groups of cognate papers in pure and applied mathematics. Papers intended for publication in the *Memoirs* should be addressed to one of the following editors. The AMS uses Centralized Manuscript Processing for initial submissions to AMS journals. Authors should follow instructions listed on the Initial Submission page found at www.ams.org/memo/memosubmit.html.

Algebra to ALEXANDER KLESHCHEV, Department of Mathematics, University of Oregon, Eugene, OR 97403-1222; email: ams@noether.uoregon.edu

Algebraic geometry and its application to MINA TEICHER, Emmy Noether Research Institute for Mathematics, Bar-Ilan University, Ramat-Gan 52900, Israel; email: teicher@macs.biu.ac.il

Algebraic geometry to DAN ABRAMOVICH, Department of Mathematics, Brown University, Box 1917, Providence, RI 02912; email: amsedit@math.brown.edu

Algebraic topology to ALEJANDRO ADEM, Department of Mathematics, University of British Columbia, Room 121, 1984 Mathematics Road, Vancouver, British Columbia, Canada V6T 1Z2; email: adem@math.ubc.ca

Combinatorics to JOHN R. STEMBRIDGE, Department of Mathematics, University of Michigan, Ann Arbor, Michigan 48109-1109; email: FRS@umich.edu

Complex analysis and harmonic analysis to ALEXANDER NAGEL, Department of Mathematics, University of Wisconsin, 480 Lincoln Drive, Madison, WI 53706-1313; email: nagel@math.wisc.edu

Differential geometry and global analysis to LISA C. JEFFREY, Department of Mathematics, University of Toronto, 100 St. George St., Toronto, ON Canada M5S 3G3; email: jeffrey@math.toronto.edu

Dynamical systems and ergodic theory and complex anaysis to YUNPING JIANG, Department of Mathematics, CUNY Queens College and Graduate Center, 65-30 Kissena Blvd., Flushing, NY 11367; email: Yunping.Jiang@qc.cuny.edu

Functional analysis and operator algebras to DIMITRI SHLYAKHTENKO, Department of Mathematics, University of California, Los Angeles, CA 90095; email: shlyakht@math.ucla.edu

Geometric analysis to WILLIAM P. MINICOZZI II, Department of Mathematics, Johns Hopkins University, 3400 N. Charles St., Baltimore, MD 21218; email: trans@math.jhu.edu

Geometric analysis to MARK FEIGHN, Math Department, Rutgers University, Newark, NJ 07102; email: feighn@andromeda.rutgers.edu

Harmonic analysis, representation theory, and Lie theory to ROBERT J. STANTON, Department of Mathematics, The Ohio State University, 231 West 18th Avenue, Columbus, OH 43210-1174; email: stanton@math.ohio-state.edu

Logic to STEFFEN LEMPP, Department of Mathematics, University of Wisconsin, 480 Lincoln Drive, Madison, Wisconsin 53706-1388; email: lempp@math.wisc.edu

Number theory to JONATHAN ROGAWSKI, Department of Mathematics, University of California, Los Angeles, CA 90095; email: jonr@math.ucla.edu

Partial differential equations to GUSTAVO PONCE, Department of Mathematics, South Hall, Room 6607, University of California, Santa Barbara, CA 93106; email: ponce@math.ucsb.edu

Partial differential equations and dynamical systems to PETER POLACIK, School of Mathematics, University of Minnesota, Minneapolis, MN 55455; email: polacik@math.umn.edu

Probability and statistics to RICHARD BASS, Department of Mathematics, University of Connecticut, Storrs, CT 06269-3009; email: bass@math.uconn.edu

Real analysis and partial differential equations to DANIEL TATARU, Department of Mathematics, University of California, Berkeley, Berkeley, CA 94720; email: tataru@math.berkeley.edu

All other communications to the editors should be addressed to the Managing Editor, ROBERT GURALNICK, Department of Mathematics, University of Southern California, Los Angeles, CA 90089-1113; email: guralnic@math.usc.edu.

Titles in This Series

913 **Ethan Akin, Joseph Auslander, and Eli Glasner,** The topological dynamics of Ellis actions, 2008

912 **Igor Chueshov and Irena Lasiecka,** Long-time behavior of second order evolution equations with nonlinear damping, 2008

911 **John Locker,** Eigenvalues and completeness for regular and simply irregular two-point differential operators, 2008

910 **Joel Friedman,** A proof of Alon's second eigenvalue conjecture and related problems, 2008

909 **Cameron McA. Gordon and Ying-Qing Wu,** Toroidal Dehn fillings on hyperbolic 3-manifolds, 2008

908 **J.-L. Waldspurger,** L'endoscopie tordue n'est pas si tordue, 2008

907 **Yuanhua Wang and Fei Xu,** Spinor genera in characteristic 2, 2008

906 **Raphaël S. Ponge,** Heisenberg calculus and spectral theory of hypoelliptic operators on Heisenberg manifolds, 2008

905 **Dominic Verity,** Complicial sets characterising the simplicial nerves of strict ω-categories, 2008

904 **William M. Goldman and Eugene Z. Xia,** Rank one Higgs bundles and representations of fundamental groups of Riemann surfaces, 2008

903 **Gail Letzter,** Invariant differential operators for quantum symmetric spaces, 2008

902 **Bertrand Toën and Gabriele Vezzosi,** Homotopical algebraic geometry II: Geometric stacks and applications, 2008

901 **Ron Donagi and Tony Pantev (with an appendix by Dmitry Arinkin),** Torus fibrations, gerbes, and duality, 2008

900 **Wolfgang Bertram,** Differential geometry, Lie groups and symmetric spaces over general base fields and rings, 2008

899 **Piotr Hajłasz, Tadeusz Iwaniec, Jan Malý, and Jani Onninen,** Weakly differentiable mappings between manifolds, 2008

898 **John Rognes,** Galois extensions of structured ring spectra/Stably dualizable groups, 2008

897 **Michael I. Ganzburg,** Limit theorems of polynomial approximation with exponential weights, 2008

896 **Michael Kapovich, Bernhard Leeb, and John J. Millson,** The generalized triangle inequalities in symmetric spaces and buildings with applications to algebra, 2008

895 **Steffen Roch,** Finite sections of band-dominated operators, 2008

894 **Martin Dindoš,** Hardy spaces and potential theory on C^1 domains in Riemannian manifolds, 2008

893 **Tadeusz Iwaniec and Gaven Martin,** The Beltrami Equation, 2008

892 **Jim Agler, John Harland, and Benjamin J. Raphael,** Classical function theory, operator dilation theory, and machine computation on multiply-connected domains, 2008

891 **John H. Hubbard and Peter Papadopol,** Newton's method applied to two quadratic equations in $\mathbb{C}^2$ viewed as a global dynamical system, 2008

890 **Steven Dale Cutkosky,** Toroidalization of dominant morphisms of 3-folds, 2007

889 **Michael Sever,** Distribution solutions of nonlinear systems of conservation laws, 2007

888 **Roger Chalkley,** Basic global relative invariants for nonlinear differential equations, 2007

887 **Charlotte Wahl,** Noncommutative Maslov index and eta-forms, 2007

For a complete list of titles in this series, visit the
AMS Bookstore at **www.ams.org/bookstore/**.